SpringerBriefs in Statistics

For further volumes:
http://www.springer.com/series/8921

Richard William Farebrother

L_1-Norm and L_∞-Norm Estimation

An Introduction to the Least Absolute Residuals, the Minimax Absolute Residual and Related Fitting Procedures

 Springer

Richard William Farebrother
The University of Manchester
Manchester
UK

ISSN 2191-544X   ISSN 2191-5458   (electronic)
ISBN 978-3-642-36299-6   ISBN 978-3-642-36300-9   (eBook)
DOI 10.1007/978-3-642-36300-9
Springer Heidelberg New York Dordrecht London

Library of Congress Control Number: 2013933582

Printed on acid-free paper

Springer is part of Springer Science+Business Media (www.springer.com)

Contents

Chapter 1
Introduction

Abstract This monograph provides an introduction to a class of linear fitting procedures that employ the sum of the absolute residuals (or L_1-norm), the minimax absolute residual (or L_∞-norm) and the median squared residual as optimality criteria in the context of the standard linear statistical model. The least absolute residuals procedure was proposed by Boscovich in 1757 and again in 1760 and discussed by Laplace, Gauss and Edgeworth; the least squares procedure was probably used by Gauss in 1794 or 1795 but first proposed in print by Legendre in 1805 before being discussed by Gauss, Laplace and many other leading scientists; finally, the minimax absolute residual procedure was proposed by Laplace in 1786, 1793 and 1799 before being discussed by Cauchy, Fourier, Chebyshev and others. The least squares and least absolute residuals procedures are widely used in statistical applications but the minimax procedure had received little support in this area until a variant, the least median of squares procedure, was proposed by Rousseeuw in 1984. Almost by definition, this last procedure is more robust to the presence of outlying observations than are the other two fitting procedures.

Keywords Rogerius Josephus Boscovich (1711–1787) · Carl Friedrich Gauss (1777–1855) · Pierre-Simon Laplace (1749–1827)

The years 2010 and 2011 respectively mark the 250th anniversary of Boscovich's constrained variant of the least sum of absolute residuals or L_1-norm line fitting procedure and the 225th anniversary of Laplace's unconstrained minimax absolute residual or L_∞-norm line fitting procedure. It therefore seems appropriate to celebrate this double anniversary by taking the opportunity of drawing the close relationship between these two fitting procedures to the attention of a wider audience.

Some practitioners might argue that, whilst the L_1-norm fitting procedure may have a peripheral role in statistics, the use of the L_∞-norm criterion (sometimes named for Chebyshev) is properly restricted to the fields of approximation theory and game theory. But this is not the case, as, in Chap. 6, we shall define a variant of the L_∞-norm procedure, known as the Least Median of Squares or LMS procedure, which incorporates the high-breakdown feature of the L_1-norm procedure and plays

R. W. Farebrother, *L_1-Norm and L_∞-Norm Estimation,*
SpringerBriefs in Statistics, DOI: 10.1007/978-3-642-36300-9_1,
© The Author(s) 2013

an increasingly significant role in the field of robust statistical analysis. For further details of this and other robust fitting procedures, see the relevant articles in the volumes edited by Dodge (1987, 1992, 1997, 2002).

The author has published three books, two of which relate to aspects of the subject covered by this brief. In his first book, Farebrother (1988a) was concerned with computational aspects of the least sum of squared residuals or L_2-norm fitting procedure and is thus of little immediate interest to readers of the present brief.

In his second book, Farebrother 1999 gives a detailed account of the history (to 1930) of the L_1-norm, L_2-norm and L_∞-norm fitting procedures. We shall therefore not address this topic in this brief but shall refer interested readers to the relevant chapters of Farebrother (1999) or to the alternative texts published by Hald (1998) and Stigler (1986) for more detailed accounts of this facet of the history of the calculus of observations. Readers may also like to consult Heyde and Seneta (2001) for brief accounts of the lives of Boscovich, Chebychev, Edgeworth, Gauss and Laplace and Stigler (1999) for related material.

Similarly, in his third book, Farebrother (2002) has given a fairly detailed account of the various geometrical and mechanical models of the L_1-norm, L_2-norm, L_∞-norm and LMS fitting procedures. Our account of mechanical models in Chap. 7 of this brief is therefore restricted to a summary of models of four variants of the L_1-norm fitting procedure supplemented by brief notes on two variants of the L_2-norm and L_∞-norm fitting procedures.

Finally, we note that, in a conventional textbook, algebraic expressions of the type discussed in this brief are usually illustrated by means of graphical figures or diagrams but, unfortunately, the author has been blind for some twenty years and is thus not able to use the relevant graphical software, so, perforce, readers will be obliged to follow Laplace's prescription by abstaining from the use of diagrams when studying these fitting procedures. Alternatively, they may prefer to read the present brief in conjunction with the relevant illustrations from Farebrother (2002).

References

Dodge, Y. (Ed.). (1987). *Statistical data analysis based on the L_1-Norm and Related Methods*. Amsterdam: North-Holland Publishing Company.

Dodge, Y. (Ed.). (1992). *L_1-Statistical analysis and related methods*. Amsterdam: North-Holland Publishing Company.

Dodge, Y. (Ed.). (1997). *L_1-Statistical procedures and related topics*. Hayward: Institute of Mathematical Statistics.

Dodge, Y. (Ed.). (2002). *Statistical data analysis based on the L_1-Norm and Related Methods*. Basel: Birkhäuser Publishing.

Farebrother, R. W. (1988a). *Linear least squares computations*. New York: Marcel Dekker.

Farebrother, R. W. (1999). *Fitting linear relationships: A history of the calculus of observations 1750–1900*. New York: Springer.

Farebrother, R. W. (2002). *Visualizing statistical models and concepts*. New York: Marcel Dekker. Available online from TaylorandFrancis.com.

Hald, A. (1998). *A history of mathematical statistics 1750–1930*. New York: Wiley.

Heyde, C. C., & Seneta, E. (Eds.). (2001). Statisticians of the centuries. Springer: New York. Reprinted by StatProb.com in 2010.

Stigler, S. M. (1986). *The history of statistics: The measurement of uncertainty before 1900.* Cambridge: Harvard University Press.

Stigler, S. M. (1999). *Statistics on the table: The history of statistical concepts and methods.* Cambridge: Harvard University Press.

Chapter 2
Point Fitting Problems in One and Two Dimensions

Abstract We begin our analysis by considering the fitting of a single point to a number of point observations in one-dimensional space. Using the L_t-norm as optimality criterion with $t = 1$, $t = 2$ or $t = \infty$, we obtain the median, mean and midrange of a set of observations respectively. Similarly, applying the same three optimality criteria in the two-dimensional case, we obtain the mediancentre or centre of population, the centroid or centre of gravity and the unnamed centre of the circle of smallest radius respectively. Moreover, if we omit some of the more extreme observations then we obtain truncated variants of these procedures. As noted in Chap. 7, the midrange and its generalisations may be associated with a set of more or less familiar geometrical instruments: The univariate midrange with a pair of callipers, the bivariate midrange with a pair of compasses and the minimax fitted line of Chap. 3 with a pair of parallel rules.

Keywords Centre of gravity · Centre of population · Centroid · L_1-norm · L_2-norm · L_∞-norm · Linear programming · Mean · Median · Midrange · Mediancentre · Oja's bivariate median · Truncated midrange

2.1 Point Fitting Problems in One Dimension

Let y_1, y_2, ..., y_n represent a set of n observations on a single variable Y, then these n observations may be represented by the n points at $y = y_1$, $y = y_2$, ..., $y = y_n$ on the y-axis of a Cartesian diagram. Moreover, we may identify a point of best fit to these n points by choosing a value for a in such a way that the sum of the squared distances

$$\sum_{i=1}^{n}(y_i - a)^2$$

is minimised.

R. W. Farebrother, L_1-*Norm and L_∞-Norm Estimation*,
SpringerBriefs in Statistics, DOI: 10.1007/978-3-642-36300-9_2,
© The Author(s) 2013

Taking square roots in this optimality function, we find that we may alternatively choose a value for a to minimise the root mean squared deviation function

$$\left[\sum_{i=1}^{n} |y_i - a|^2 \right]^{\frac{1}{2}}.$$

and, replacing 2 by p with $0 < p \le \infty$, in this expression, we have the optimality criterion employed in the more general L_t-norm point fitting problem

$$\left[\sum_{i=1}^{n} |y_i - a|^t \right]^{\frac{1}{t}}.$$

In this brief, we shall be largely concerned with two special cases of the L_t-norm problem: the first is identified by setting $t = 1$, when we have the sum of absolute deviations optimality criterion employed in the L_1-norm point fitting problem:

$$\sum_{i=1}^{n} |y_i - a|$$

and, in the limit as p tends to ∞, we have the minimax absolute residual optimality criterion employed in the corresponding L_∞-norm point fitting problem:

$$max_{i=1}^{n} |y_i - a|.$$

Both of these special cases of the general point fitting problem are readily solved: the L_t-norm optimality criterion defines the median (or middlemost) observation when $t = 1$, the arithmetic mean of the observations $\bar{y} = \sum_{i=1}^{n} y_i/n$ when $t = 2$, and the midrange (or midpoint of the shortest line segment containing all n observations) in the limit as t tends to ∞.

To define the median and the conventional midrange of the n observations, $y_1, y_2, \ldots, y_n$, we arranged these observations in increasing order as $y_{[1]} \le y_{[2]} \le \ldots \le y_{[n]}$, then the median value of these n observations is given by $(y_{[m]} + y_{[m]})/2 = y_{[m]}$ when $n = 2m - 1$ is odd and by $(y_{[m]} + y_{[m+1]})/2$ when $n = 2m$ is even. Similarly, the conventional midrange is given by $(y_{[1]} + y_{[n]})/2$.

Now, all three of these expressions take the form $(y_{[r+1]} + y_{[n-r]})/2$ where $0 \le r \le n/2$, which we shall call the (r, r)-level symmetrically truncated midrange as its computation ignores the r smallest values and the r largest values of y_i.

As a more general variant of this expression, we may define the (r, s)-level non-symmetrically truncated midrange $(y_{[r+1]} + y_{[n-s]})/2$ whose computation ignores the r smallest values and the s largest values of y_i. In particular, if we wish to retain m observations in the nonsymmetric case, then we have to choose values for r and s in such a way as to exclude $r + s = n - m$ observations from the computation so that the remaining $n - r - s = m$ observations define the midpoint of the shortest line segment covering m of the observations. Moreover, if $n = 2m$ or $n = 2m - 1$ then this expression excludes one-half (or almost one-half) of the observations from the computation, and the optimal nonsymmetric truncated midrange is known as the 'shortest half'.

2.2 Point Fitting Problems in Two Dimensions

Generalising the representation of Sect. 2.1 to the 2-dimensional case, we find that we have n observations on the two variables X and Y. Let (x_1, y_1), (x_2, y_2), ..., (x_n, y_n) represent a set of n matched pairs of observations on the two variables X and Y, then, for $i = 1, 2, \ldots, n$, the ith observation may be represented by a point at $(x, y) = (x_i, y_i)$ in the xy-plane of a two-dimensional Cartesian diagram.

In this context, and for each choice of $t > 0$, two definitions of the point of best fit become available: we may either separately minimise the L_t-norm of the distances measured perpendicular to the x-axis (and thus parallel to the y-axis)

$$\left[\sum_{i=1}^{n} |x_i - c|^t \right]^{\frac{1}{t}}$$

to obtain an optimal value for c at the same time as minimising the L_t-norm of the absolute distances measured perpendicular to the y-axis (and thus parallel to the x-axis)

$$\left[\sum_{i=1}^{n} |y_i - a|^t \right]^{\frac{1}{t}}$$

to obtain an optimal value for a.

Alternatively, we may simultaneously choose values for a and c to minimise the L_t-norm of the n Euclidean distances

$$\left\{ \sum_{i=1}^{n} \left[(x_i - c)^2 + (y_i - a)^2 \right] \right\}^{\frac{1}{t}}.$$

In the special case when $t = 2$ the square of this last expression may be written as

$$\left[\sum_{i=1}^{n} (x_i - c)^2 \right] + \left[\sum_{i=1}^{n} (y_i - a)^2 \right]$$

so that we obtain the same values $c = \bar{x}$ and $a = \bar{y}$ in this context as in the componentwise case mentioned above, where $\bar{y} = \sum_{i=1}^{n} y_i / n$ and $\bar{x} = \sum_{i=1}^{n} x_i / n$.

Thus, this alternative expression defines the mediancentre or centre of population of the n observations when $t = 1$, the centroid or centre of gravity when $t = 2$, and the unnamed centre of the circle of smallest area (and hence smallest radius) which just covers all n points in the limit as t tends to ∞.

As in Sect. 2.1, and for all values of $m \leq n$, we may readily generalise our definition of the one-dimensional nonsymmetrically truncated midrange to the two-dimensional case by replacing the midpoint of the shortest line segment which just covers $m \leq n$ points by the centre of the circle with smallest area which just covers $m \leq n$ points. Indeed, and for all values of $p \geq 1$, this last definition may be further generalised to the centre of the p-dimensional sphere with minimal p-dimensional volume which just covers $m \leq n$ points.

2.3 Truncated Point Fitting Problems in Two Dimensions

The mediancentre of a set of two-dimensional observations is a point (x_0, y_0) or (c, a) chosen in such a way that the sum of the lengths (or Euclidean distances) of the line segments joining the n given points to this arbitrary point takes its minimum value. [Gower (1974) has supplied an algorithm for performing the necessary calculations.]

Now, the line segments in this definition of the mediancentre may be replaced by triangles, triangular pyramids, ..., that is, by p-dimensional simplices where $p = 2$, 3, ... Thus an alternative definition of a central point of a set of two-dimensional observations is a point (x_0, y_0) chosen in such a way as to minimise the sum of the areas of the $^nC_2 = n(n-1)/2$ triangles defined by any two of the n given points and the arbitrary point. The centre defined in this way is known as Oja's (1983) bivariate median.

Now, for all $i < j$, the area of the triangle with vertices (x_0, y_0), (x_i, y_i) and (x_j, y_j) is given by one-half of the absolute value of the determinant of the 3×3 matrix

$$\begin{bmatrix} 1 & x_0 & y_0 \\ 1 & x_i & y_i \\ 1 & x_j & y_j \end{bmatrix}$$

that is, by one-half of the absolute value of

$$(x_i y_j - x_j y_i) - (y_j - y_i)x_0 + (x_j - x_i)y_0$$

or one-half of $w_{ij}|e_{ij}|$ where

$$w_{ij} = |x_j - x_i|$$
$$e_{ij} = y_0 - a_{ij} - x_0 b_{ij}$$
$$a_{ij} = (x_j y_i - x_i y_j)/(x_j - x_i)$$
$$b_{ij} = (y_j - y_i)/(x_j - x_i)$$

and our problem takes the form of a weighted least absolute residuals fitting problem of the type discussed in Sect. 3.4 below provided that the coefficient of y_0 is nonzero, that is, provided that the x_i are distinct.

In other words, we have to choose x_0 and y_0 in such a way as to minimise a weighted sum of absolute values. Thus, the procedure for determining the value of Oja's bivariate median may be implemented in the form of a linear programming problem. Niinimaa et al. (1992) have provided such an algorithm.

The concept underlying Oja's bivariate median may readily be generalised to higher dimensions if we choose to minimise the sum of the p-dimensional volumes of the ^nC_p distinct p-dimensional simplices defined by the arbitrary point and any set of p of the n given points. In this case, we have to minimise the sum of the absolute values of the determinants of a set of $(p + 1) \times (P + 1)$ matrices divided by $p!$, see Farebrother (1992) for details.

References

Dodge, Y. (Ed.). (1992). L_1-Statistical analysis and related methods. Amsterdam: North-Holland Publishing Company.

Farebrother, R. W. (1992). The geometrical foundations of a class of estimation procedures which minimise sums of Euclidean distances and related quantities, in Dodge (pp. 337–349).

Gower, J. C. (1974). Mediancentre. Applied Statistics, 23, 466–470.

Niinimaa, A., Oja, H., & Nyblom, J. (1992). The Oja bivariate median. Applied Statistics, 41, 611–617.

Oja, H. (1983). Descriptive statistics for multivariate distributions. Statistics and Probability Letters, 1, 327–332.

Chapter 3
The Hyperplane Fitting Problem
in Two or More Dimensions

Abstract We continue our analysis by considering the fitting of a p-dimensional hyperplane to a set of point observations in $(p + 1)$-dimensional space. Again using the L_t-norm as optimality criterion with $t = 1, t = 2$ or $t = \infty$, we obtain the least absolute residuals, least squared residuals and minimax absolute residual procedures for the fitting of the hyperplane to this set of observations. Expressing the L_t-norm problem in matrix form, we establish that the weighted L_1-norm problem is intimately associated with a *transformation* of the weighted L_∞-norm problem and *vice versa*. Then, examining the matrix representation of the L_1-norm and L_∞-norm problems, we identify a particular vector as the Lagrange multipliers of these problems. Finally, we define the corresponding matrix expression in the L_2-norm case and identify it as the formulation of the least squares problem employed in continuum regression analysis.

Keywords Hyperplane fitting problem · Line fitting problem · L_1-norm · L_2-norm · L_∞-norm · Matrix representation · Method of least squared residuals · Method of least absolute residuals · Method of minimax absolute residual

3.1 Line Fitting Problems in Two Dimensions

Generalising the problem of Chap. 2, we suppose that, instead of a point of best fit, we wish to determine a line of best fit to this set of n observations in 2-dimensional space, then we have to determine values for the unknown parameters a and b in such a way that the L_t-norm of the distances from the n points to the fitted line is minimised.

Now, these distances may either be measured perpendicular to the x-axis or perpendicular to the fitted line. In the first (conventional) case we have to choose values for a and b to minimise the L_t-norm of the Euclidean distances

R. W. Farebrother, *L_1-Norm and L_∞-Norm Estimation*,
SpringerBriefs in Statistics, DOI: 10.1007/978-3-642-36300-9_3,

$$\left[\sum_{i=1}^{n} |y_i - a - bx_i|^t\right]^{\frac{1}{t}}.$$

Similarly, in the second (orthogonal) case, we have to choose values for a and b to minimise the scaled function

$$\left\{\left[\sum_{i=1}^{n} |y_i - a - bx_i|^t\right]^{\frac{1}{t}}\right\} \bigg/ \sqrt{(1 + b^2)}$$

where all terms under the summation sign have been divided by a common factor of $\sqrt{(1 + b^2)} \geq 1$ to reflect the reduction in distances between the n given points and the fitted line when the direction in which measurements are to be taken is rotated from a line lying perpendicular to the x-axis to one lying perpendicular to the fitted line.

Most readers will already be familiar with these optimisation criteria when $t = 2$. In this context, the first criterion defines the conventional least squares line fitting procedure whilst the latter serves as the basis of the method of orthogonal least squares and leads on to the development of a set of (orthogonal) least squares principal components. Unfortunately, any discussion of L_1-norm and L_∞-norm variants of these orthogonal procedures is well beyond the scope of the present brief, but see Nyquist (1988, 2002) for an introduction.

Indeed, in this brief we shall be largely concerned with two special cases of the conventional (unweighted) L_t-norm line fitting problem: the first of which is identified by setting $t = 1$, when we have the optimality criterion employed in the conventional unweighted L_1-norm line fitting problem:

$$\sum_{i=1}^{n} |y_i - a - bx_i|$$

and, in the limit as p tends to ∞, we have the optimality criterion employed in the conventional unweighted L_∞-norm line fitting problem:

$$max_{i=1}^{n} |y_i - a - bx_i|.$$

Moreover, we have already solved both of these problems in Sect. 2.1 in the special case when the slope parameter b is set equal to zero.

3.2 Hyperplane Fitting Problems in Higher Dimensions

Thus far in this brief, and for ease of exposition, we have restricted our attention to the conventional fitting of a zero-dimensional point to n observations in one-dimensional or two-dimensional space and to the conventional fitting of a one-dimensional line to n observations in two-dimensional space. In the remainder of the present Chapter and in Chaps. 4–6 we generalise our analysis to the p-dimensional case with $p \geq 1$. That is, given a value for $p \geq 1$, we address the more general problem of fitting a p-dimensional hyperplane to n observations in $(p + 1)$-dimensional space which incorporates the problem of fitting a $(p-1)$-dimensional hyperplane to n observations in p-dimensional space if all observations on X_1 take unit values.

For $i = 1, 2, ..., n$, let $\{x_{i1}, x_{i2}, ..., x_{ip}, y_i\}$ represent the ith observation on a set of $p+1$ variables $\{X_1, X_2, ..., X_p, Y\}$. Then, for $t > 0$, the conventional unweighted L_t-norm hyperplane fitting procedure chooses values for $b_1, b_2, ..., b_p$ to minimise the L_t-norm of the residuals

$$\|e\|_t = \left[\sum_{i=1}^{n} |e_i|^t \right]^{\frac{1}{t}}$$

where, for $i = 1, 2, ..., n$, the ith residual is defined by

$$e_i = y_i - x_{i1}b_1 - x_{i2}b_2 - ... - x_{ip}b_p.$$

As noted in Sect. 3.1, the most familiar L_t-norm fitting procedure, known as the least squares procedure, sets $t = 2$ and chooses values for $b_1, b_2, ..., b_p$ to minimise the sum of the squared residuals $\sum_{i=1}^{n} e_i^2$. A second choice, known as the least absolute residuals procedure, sets $t = 1$ and chooses $b_1, b_2, ..., b_p$ to minimise the sum of the absolute residuals $\sum_{i=1}^{n} |e_i|$ and a third choice, known as the minimax absolute residual procedure, sets $t = \infty$ and chooses $b_1, b_2, ..., b_p$ to minimise the largest absolute residual $max_{i=1}^{n} |e_i|$.

3.3 Matrix Representation of the Problem

For economy of notation, we now translate the above discussion of the general L_t-norm fitting problem into matrix form. We consider the fitted model

$$y = Xb + e$$

where y is an $n \times 1$ matrix of observations on the dependent variable, X is an $n \times p$ matrix of observations on the p explanatory variables, b is a $p \times 1$ matrix of fitted values and $e = y - Xb$ is an $n \times 1$ matrix of residuals. For ease of exposition, we shall assume that the $n \times p$ matrix X has full column rank p.

Let t be a fixed value in the range $0 < t \leq \infty$, then b is said to be a (unweighted) L_t-norm fitting procedure if it chooses b to minimise the optimality function

$$\|e\|_t \;=\; \left[\sum_{i=1}^{n} |e_i|^t\right]^{\frac{1}{t}}.$$

As noted above, the most familiar members of this class of fitting procedures are the L_1-norm fitting procedure which minimises the sum of the absolute residuals, $\sum |e_i|$, the L_2-norm fitting procedure which minimises the sum of the squared (absolute) residuals, $\sum e_i^2$, and the L_∞-norm fitting procedure which minimises the largest absolute residual, $max|e_i|$.

Let w be a given $n \times 1$ matrix of positive weights. Then these unweighted fitting procedures may be generalised to yield the corresponding weighted L_1-norm fitting procedure which minimises the weighted sum of the absolute residuals, $\sum w_i|e_i|$, the weighted L_2-norm fitting procedure which minimises the weighted sum of the squared (absolute) residuals, $\sum w_i e_i^2$, and the weighted L_∞-norm fitting procedure which minimises the largest weighted absolute residual, $max w_i|e_i|$.

3.4 The Weighted L_1-Norm and L_∞-Norm Fitting Problems

A further slight generalisation of the L_1-norm problem of Sect. 3.3 replaces the common value of w by two possibly distinct $n \times 1$ nonzero matrices w_A and w_B with nonnegative elements. In this context, we suppose that we are given an $n \times p$ matrix X of rank p, and an $n \times 1$ matrix y that is not linearly dependent on the columns of X. Further, let e_A and e_B be $n \times 1$ nonnegative matrices such that $e = e_A - e_B$, then our slight generalisation of the weighted L_1-norm fitting problem of Sect. 3.3 may be written as:

> Problem P_1:
>
> Minimise $j = w_A' e_A + w_B' e_B$
>
> subject to
>
> $\quad Xb + e_A - e_B = y$
>
> and $e_A \geq 0, e_B \geq 0$.

In the familiar unweighted L_1-norm problem all the elements of w_A and w_B take unit values. However, in the special case of regression quantiles considered by Koenker and Bassett (1978), all the elements of w_A take the same nonnegative value and all the elements of w_B take the same (if usually distinct) nonnegative value.

If we now define T an $n \times q$ matrix of rank $q = n - p - 1$ satisfying $T'X = 0$ and $T'y = 0$, s a nonzero $n \times 1$ matrix of rank 1 satisfying $s'X = 0$ and $s'y = m > 0$, and Q an $n \times p$ matrix of rank p such that the $n \times n$ matrix $[Q\, s\, T]$ is nonsingular. Then problem P_1 may be written as

Problem P_2:

Minimise $j = w'_A e_A + w'_B e_B$

subject to

$$T' e_A - T' e_B = 0$$
$$s' e_A - s' e_B = m$$
$$Q' X b + Q' e_A - Q' e_B = Q' y$$

and $e_A \geq 0, e_B \geq 0$

or as

Problem P_3:

Minimise $j = w'_A e_A + w'_B e_B$

subject to

$$T' e_A - T' e_B = 0$$
$$s' e_A - s' e_B = m$$

and $e_A \geq 0, e_B \geq 0$

before forming

$$b^* = (Q' X)^{-1} Q' (y - e_A^* + e_B^*)$$

where the last of these equations expresses the optimal value of b in terms of the optimal values of e_A and e_B.

Supposing that the corresponding optimal value of j is strictly positive (as it usually will be), we may set $e_A = j f_A$ and $e_B = j f_B$ to obtain

Problem P_4:

Maximise $m/j = s' f_A - s' f_B$

subject to

$$T' f_A - T' f_B = 0$$
$$w'_A f_A + w'_B f_B = 1$$

and $f_A \geq 0, f_B \geq 0$.

Replacing each of the equalities in Problems P_1–P_4 by a matched pair of $\leq$ and $\geq$ inequalities and replacing each unconstrained variable by a difference between two nonnegative variables, we find that each of Problems P_1–P_4 take the form of standard linear programming problems and deduce that the linear programming dual of Problem P_4 is given by

Problem P_5:

Minimise k

subject to

$$g = s - Td$$
$$\text{and} \quad -kw_A \leq g \leq kw_B.$$

Thus, we have established that the weighted L_1-norm problem P_1 is intimately associated with a transformation of the weighted L_∞-norm problem P_5.

We now consider a similar slight variant of the weighted L_∞-norm problem of Sect. 3.3 which chooses a value for b to minimise $max(w_i|e_i|)$ where $e = y - Xb$. For expository purposes, it is convenient to change notation and replace p, y, X, b and e by q, s, T, d and g and to replace the ith weight w_i by its reciprocal $1/w_i$. In this context, we may suppose that we are given an $n \times q$ matrix T of rank q, and an $n \times 1$ nonzero matrix s that is not linearly dependent on the columns of T. Then, in this revised notation, our slight generalisation of the weighted L_∞-norm fitting problem chooses d to minimise $max|g_i|/w_i$ which may be written as problem P_5 above. Where w_A and w_B are again $n \times 1$ nonzero matrices with nonnegative elements, In particular, all the elements of w_A and w_B take unit values in the familiar unweighted L_∞-norm problem. [For a statement of the weighted L_∞-norm problem in its original notation, see Problem $P_5^{\#}$ in Sect. 5.5.]

If we now define X, an $n \times p$ matrix of rank $p = n - q - 1$ satisfying $X'T = 0$ and $X's = 0$, y, a nonzero $n \times 1$ matrix of rank 1 satisfying $y'T = 0$ and $y's = m > 0$, and Z, an $n \times q$ matrix of rank q such that the $n \times n$ matrix $[XyZ]$ is nonsingular. Then our slight generalisation of the weighted L_∞-norm fitting problem may be written as Problem P_5 above, or as

Problem P_6:

Minimise k

subject to

$$X'g = 0$$
$$y'g = m$$
$$Z'g = Z's - Z'Td$$
$$\text{and} \quad -kw_A \leq g \leq kw_B$$

or as

Problem P_7:

Minimise k

subject to

$$X'g = 0$$
$$y'g = m$$

and $-kw_A \leq g \leq kw_B$

before forming

$$d^* = (Z'T)^{-1}Z'(s - g^*)$$

where the last of these equations again expresses the optimal value of d in terms of the optimal value of g.

Supposing that the corresponding optimal value of k is strictly positive (as it usually will be), we may set $g = kh$ to obtain

Problem P_8:

Maximise $m/k = y'h$

subject to

$$X'h = 0$$

and $-w_A \leq h \leq w_B$.

Again, replacing each of the equalities in Problems P_5–P_8 by a matched pair of $\leq$ and $\geq$ inequalities and replacing each unconstrained variable by a difference between two nonnegative variables, we find that each of Problems P_5–P_8 take the form of standard linear programming problems and deduce that the linear programming dual of Problem P_8 is given by Problem P_1 or

Problem P_9:

Minimise $j = w'_A e_A + w'_B e_B$

subject to

$$Xb + e_A - e_B = y$$

and $e_A \geq 0, e_B \geq 0$

Thus, we have established that the weighted L_∞-norm problem P_5 is intimately associated with a transformation of the weighted L_1-norm problem $P_9 = P_1$ and deduce that either of these problems may be solved directly or indirectly by solving any one of the eight linear programming problems identified above.

Indeed, Farebrother (2002) has used these relationships to convert Laplace's (1786) original L_∞-norm fitting problem with $n = 4$ observations on $q = 2$ explanatory variables into an equivalent L_1-norm fitting problem with $n = 4$ observations on $p = 1$ explanatory variable, noting that the latter problem is easily solved. [For related work on the L_1-norm fitting problem, see Seneta and Steiger (1984).]

3.5 Relation to the L_2-Norm Fitting Problem

Before embarking on our account of the computational procedures to be employed in the solution of Problems P_1–P_9, we must discuss an L_2-norm variant of Problems P_4 and P_8. On setting $w_A = w_B$ equal to an $n \times 1$ column of ones and $f = f_A - f_B$, we find that problem P_4 may be rewritten as:

> Problem $P_4^\#$:
>
> Maximise $m/j = s'f$
>
> subject to
>
> $$T'f = 0$$
>
> and $\|f\|_1 = 1$

where $\|f\|_1 = \sum_{i=1}^{n} |f_i|$, so that this revised problem takes the same form as problem P_8 except that the former employs the L_1-norm condition $\|f\|_1 = 1$ whilst the latter employs the L_∞-norm condition $\|h\|_\infty = 1$.

This similarity between these two variants of problems P_4 and P_8 prompts us to examine the corresponding L_2-norm problem:

> Problem P_{10}:
>
> Maximise $y'h$
>
> subject to
>
> $$X'h = 0$$
>
> and $\|h\|_2 = 1$

where $\|h\|_2 = \sqrt{(\sum_{i=1}^{n} h_i^2)}$. This constrained function is clearly optimised by setting h equal to $h^* = y^*/r$ where y^* is the $n \times 1$ matrix of least squares residuals $y^* = y - X(X'X)^{-1}X'y$ and $r = \|y^*\|_2 = \sqrt{(y^{*\prime}y^*)}$.

Further, let Z_o be an $n \times (n - p)$ matrix of rank $n - p$ satisfying $Z_o'X = 0$, then $X'h = 0$ implies that $h = Z_o c$ for some $(n - p) \times 1$ matrix c, and, on squaring the (nonnegative) objective function, we have the problem

> Problem P_{10}^*:
>
> Maximise $c'Z_o'yy'Z_o c$
>
> subject to
>
> $$c'Z_o'Z_o c = 1$$

which is a variant of the characterisation of the normalised conventional least squares fitting procedure $c^* = (Z_o'Z_o)^{-1}Z_o'y/r$ employed in continuum regression analysis by Stone and Brooks (1990) and Sundberg (1993) where $r^2 = y'My$ and

$$M = Z_o(Z_o'Z_o)^{-1}Z_o' = I_n - X(X'X)^{-1}X'$$

In this context, $h^* = Z_o c^* = My/r$ is clearly a scalar multiple of the least squares residuals $e^* = My$.

Finally, we note that f in Problem P_4 and h in Problem P_8 may each be identified as the $n \times 1$ matrices of Lagrange multipliers associated with the solution of the L_1-norm fitting problem P_1 and the L_∞-norm fitting problem P_5 respectively.

References

Dodge, Y. (Ed.). (2002). *Statistical Data Analysis based on the L_1-Norm and Related Methods*. Basel: Birkhäuser Publishing.

Farebrother, R. W. (2002). A dual formulation of Laplace's minimax problem. *Student, 4*, 81–85.

Koenker, R., & Bassett, G. W. (1978). Regression quantiles. *Econometrica, 46*, 33–50.

Laplace P. S. (1786). Mémoire sur la figure de la terre, Mémoires de l'Académie Royale des Sciences de Paris [pour 1783], (pp. 17–46). Reprinted in his Oeuvres Complètes (Vol. 1, pp. 3–32). Paris: Gauthier-Villars, 1895.

Nyquist, H. (1988). Least orthogonal absolute deviations. *Computational Statistics and Data Analysis, 6*, 361–367.

Nyquist H. (2002). Orthogonal L_1-norm estimation, in Dodge (pp. 171–182).

Seneta, E., & Steiger, W. L. (1984). A new LAD curve-fitting algorithm: Slightly overdetermined equation systems. *Discrete Applied Mathematics, 7*, 79–91.

Stone, M., & Brooks, R. J. (1990). Continuum regression: cross-validated sequentially constructed prediction embracing ordinary least squares, partial least squares and principal components regression. *Journal of the Royal Statistical Society (Series B), 52*, 237–269.

Sundberg, R. (1993). Continuum regression and ridge regression. *Journal of the Royal Statistical Society (Series B), 55*, 653–659.

Chapter 4
Linear Programming Computations

Abstract After sketching the graphical solution of the L_1 -norm and L_∞-norm fitting problems based on a plot of lines in parameter space, we survey the pertinent literature on modern standard and improved linear programming solutions to these problems. We investigate the possibility that a variant of the familiar simplex procedure could have been developed some thirty years before it actually appeared in the late 1940s. Finally, we survey a range of possible alternatives to using the L_t-norm procedure in the limit as t tends to 1 or ∞ as possible practical solutions to the problem of non-uniqueness of the solution to the L_1-norm and L_∞-norm procedures.

Keywords Computational procedures · Euclidean geometry · Graphical solution · Linear programming · Linear programming duality · Non-uniqueness · Primal and dual simplex · Projective geometry · Projective geometry duality · Charles-Jean-Gustave-Nicolas de la Vallée Poussin (1866–1962) · Francis Ysidro Edgeworth (1845–1926) · Jean Baptiste Joseph Fourier (1768–1830).

4.1 Regular and Projective Geometry Solutions of the L_1-Norm Problem

Nowadays, it is well-known that the solution of the unweighted L_1-norm fitting problem of Sect. 3.2 is characterised by a set of p or more zero residuals and that the solution of the unweighted L_∞-norm fitting problem is characterised by a set of $p + 1$ or more residuals of the same absolute size (not all of which take the same sign). The first of these conditions was first stated by Gauss (for the general case $p \geq 1$) in 1809 and the second by Laplace (for the special case $p = 2$) in 1786.

Setting the $n \times 1$ matrices w_A and w_B equal to columns of ones in Sect. 3.4, we find that the solution of the unweighted L_1-norm fitting Problem P_1 is characterised by a set of p or more zero residuals whilst the solution of the unweighted L_∞-norm fitting problem P_5 is characterised by a set of $q + 1 = n - p$ or more residuals of the same absolute size (not all of which take the same sign). [These two sets of optimality conditions are clearly closely related as noted by Farebrother (2002b).]

Temporarily restricting our analysis to the case of $p = 2$ unknown parameters, we know that the solution of the unweighted L_1-norm fitting problem is characterised by a set of two zero residuals $e_i = 0$ and $e_j = 0$. Thus, any fitted line satisfying these two conditions must pass through the points (x_i, y_i) and (x_j, y_j). We thus have $n(n - 1)/2$ pairs of points (x_i, y_i) and (x_j, y_j) in xy-observation space and thus $n(n - 1)/2$ lines passing through any pair of such points from which to make our choice.

An alternative geometrical exposition of this solution procedure may be made in the context of the dual space of parameters. This form of duality, known as projective geometry duality, is clearly distinct from linear programming duality employed in Sect. 3.4 above. In this context, the observation formerly associated with the point (x_i, y_i) in xy -observation space is also associated with the equation $e_i = 0$ or $y_i = a + bx_i$ and thus with the line $a = y_i - bx_i$ in dual ba-parameter space. In a similar way, the (fitted) line $y = a + bx$ in xy -observation space is associated with the point (b, a) in ba-parameter space, see Farebrother (1989, 2002) for further details.

These two statements define a familiar duality transformation, and one which has been used implicitly or explicitly by statisticians for more than 200 years. However, this is not the only representation of the projective geometry transformation as some authors including Souvaine and Steele (1987) and Owen and Shiau (1988) prefer to employ a variant which negates the coefficient of b.

Returning to our analysis of the solution of the unweighted L_1-norm problem characterised by a set of two zero residuals $e_i = 0$ and $e_j = 0$. In the present (dual) context, we find that we have n lines $e_i = 0$ and $n(n - 1)/2$ points of intersection between pairs of these n lines. Thus, as noted by Edgeworth (1887a, b, 1888), in ba-parameter space, our solution procedure passes from point of intersection to point of intersection until the objective function $\sum |e_i|$ takes its minimal value. Although mentioned by Schwartz (1989) and Bloomfield and Steiger (1983), this simple fitting procedure was not implemented by Gauss in 1809 but Boscovich had developed a geometrical procedure in 1760 which Laplace cast in algebraic form in 1793. Both of these procedures artificially imposed the requirement that the fitted line passed through the centroid of the observations $(\bar{x}, \bar{y})$ equivalent to imposing the adding-up condition $\sum e_i = 0$ on their solution procedure. Gauss removed this unnecessary constraint in 1809 but no further progress was made with the practical solution of this problem until Edgeworth (1887a,b, 1888), latterly under the influence of Turner (1887), developed a simple procedure which Rhodes (1930) further simplified. Moreover, Farebrother (1988, 1992) has found that his implementation of Rhodes's (1930) algorithm with $p = 2$ unknowns can yield a more efficient fitting procedure than some of the better known modern algorithms mentioned in Sect. 4.3 below.

4.2 Projective Geometry Solution of the L_∞-Norm Problem

Turning our attention to the unweighted L_∞-norm fitting problem of Sect. 3.2 when there are $p = 2$ unknown parameters, and noting that the solution of this problem is characterised by a set of three residuals $\{e_i, e_j, e_k\}$ of the same absolute size, two of which take one sign whilst the third takes the opposite sign. Consider the same n observations represented by the same n residuals e_i or $a = y_i - bx_i$ in ba-parameter space. Now, we have $n(n-1)(n-2)/6$ distinct sets of three residuals $\{e_i, e_j, e_k\}$, one of which must be the negative of the other two. Thus, each set of three residuals $\{e_i, e_j, e_k\}$ yields three points in ba-parameter space defined by the double equalities $-e_i = e_j = e_k$, $e_i = -e_j = e_k$ and $e_i = e_j = -e_k$, which may represent the optimal solution of an L_∞-norm problem. Thus, as described by Fourier (1827), we again have to proceed from point to point in ba-parameter space until the objective function $max|e_i|$ takes its minimal value. Laplace (1793, 1799, 1812), de Prony (1804), Cauchy (1824, 1831), Fourier (1827) and de la Vallée Poussin (1911) have all discussed possible solution procedures, see Farebrother (1987, 1997, 1999) for details. but, so far as the author is aware, no solution procedure based on any of these algorithms have been developed as a modern computer program.

In this contest, we refer interested readers to Franksen (1985a), Grattan-Guinness (1970) and Williams (1986) for further details of Fourier's contributions to linear programming and to Franksen (1985b) and Brentjes (1994) for detailed accounts of the early history of nonlinear programming and the so-called Kuhn-Tucker theorem.

4.3 Linear Programming Computations

As noted above, all eight of the problems discussed in Sect. 3.4 may be expressed in the form of standard linear programming problems. We may therefore obtain a solution by applying any one of the standard primal or dual simplex algorithms to the chosen formulation of the problem.

On the other hand, more efficient algorithms may be obtained by incorporating the special features reflecting the particular structure of the L_1-norm and L_∞-norm fitting problems in the relevant solution procedures. The relations between the various standard and improved linear programming algorithms has been ably summarised by Arthanari and Dodge (1981) and by Bloomfield and Steiger (1983) and their relative merits discussed in detail by Watson (2000).

The best known algorithms for the general L_1-norm problem are Barrodale and Roberts (1970, 1974), Bartels et al. (1978), (Bloomfield, 1980), and the interior method of Karmarkar (1984) which came too late to be included in either of the first two surveys, but L_1-norm implementations of Karmarkar's interior method have been analysed by Koenker (1997) and Portnoy and Koenker (1997). For an analysis of the same algorithm in the context of the corresponding quantile regression problem, see Portnoy (1997).

However, as noted by Koenker (1997, pp. 16–17)

With the advent of George Dantzig's simplex algorithm in the late 1940's this situation changed dramatically, and by the mid-50's there were several formulations of the L_1 estimator for regression as a linear program and explicit simplex-based programs to compute it. The paper by Wagner (1959) clarified the important role of the L_1 dual problem. These efforts culminated in the algorithm of Barrodale and Roberts (1974) which still serves as the L_1 algorithm of choice for most statistical computing environments. Contrary to a plethora of dire warnings throughout the literature, about the difficulty of L_1 computation this algorithm actually delivers least absolute error regression estimates faster than the corresponding least squares algorithms in many packages, including *Splus* and *Stata*, for problems of moderate size, up to a few hundred observations. However, for larger problems the Barrodale and Roberts algorithm exhibits $O(n^2)$ growth in execution time and thus quickly lives up to its slothful reputation.

Although Watson's (2000) comparison of the various computer programs was comprehensive in scope, he did not feel able to come to any very firm conclusions as to which algorithm was best. In particular, Watson (2000, p. 8) has noted that:

Numerical experiments were reported by Osborne and Watson in 1996, where the secant-based method was seen to be as good as fast median methods on randomly generated problems, and to perform considerably better on problems with systematic data. Comparisons of other types of method with simplex methods really need to take this into account before definitive conclusions can be drawn.

4.4 A Hypothetical General Procedure

As noted by Farebrother (1987, 1997, 1999), procedures for solving constrained or unconstrained variants of the weighted or unweighted L_1 -norm problem were proposed by Boscovich in his notes to Stay (1760), Laplace (1793, 1799, 1812, 1818), Gauss (1809), Edgeworth (1887a,b, 1888, 1923), Bowley (1902, 1928), Rhodes (1930) and Singleton (1940), and procedures for solving the unconstrained and unweighted L_∞-norm problem were proposed by Laplace (1786, 1793, 1799, 1812), de Prony (1804), Cauchy (1824, 1831), Fourier (1827), and de la Vallée Poussin (1911).

In this connection, it is pertinent to note that, contrary to Grattan-Guinness's (1994, p. 46) assertion, Laplace did not impose the adding-up constraint on the unweighted L_∞-norm fitting procedure. Indeed, the author does not know of any early proposal to this effect.

Despite the fact that these early L_1-norm and L_∞-norm procedures exhibited many of the now-familiar features of linear programming problems, the relevant details do not seem to have been rediscovered until too late for them to make any significant contribution to the development of linear programming theory. Indeed, these historical procedures do not seem to have warranted citation in the mainstream literature of linear programming until after the publication of Dantzig's (1963) treatise on the subject.

In this connection, it seems reasonable to suggest that a careful analysis of any one of these historical approaches to the solution of the unweighted L_1-norm or L_∞-norm fitting problems by a careful researcher could have lead to the development of a general (or, at least, a more general) variant of the simplex algorithm some thirty years before it was actually proposed by Dantzig in the late 1940s. More precisely, it seems reasonable to conjecture that an early variant of the simplex procedure could have been developed at any time after the publication of de la Vallée Poussin's L_∞-norm procedure in 1911. Indeed, Farebrother (2006) has suggested a possible route by which this end might have been achieved on the initial assumption that, after suitable transformations, the problem of interest could be expressed in the form of a linear programming problem concerned with the minimisation of a linear objective function subject to a set of linear inequality constraints of the same type ($\leq$ or $\geq$) and that the coefficients attached to one of the variables in the objective function and in the inequality constraints are all strictly positive. Of course, these initial specifications would have been relaxed as practitioners became familiar with the workings of the hypothetical fitting procedure.

Grattan-Guinness (1994, p. 69) has suggested that the principal reason that practitioners made little progress in this direction may simply have been because civil and military engineers were not interested in the solution of linear optimisation problems of the type employed in linear programming analysis much before the outbreak of the Second World War. On this point, also see Brentjes (1994).

Once the simplex algorithm had been brought to the attention of practitioners, there was a burgeoning activity in the statistical literature represented by the publication of articles by Harris (1950), Charnes et al. (1955), Karst (1958), Wagner (1959), Stiefel (1959, 1960), Dolby (1960) and Fisher (1961), many of which reveal a deplorable lack of awareness of the prehistory of their subject.

There was also a parallel interest in linear programming theory in the field of economics associated with the names of Koopmans and Kantorovich, who were awarded the Nobel Prize in Economics in 1975. Much to the annoyance of Koopmans, the name of Dantzig was omitted from the prize citation as he was not an economist. see Charnes and Cooper (1962) for an early account of the contributions to linear programming in economics made by Koopmans and Kantorovich.

Some years later, Dorfman (1984), one of the authors of Dorfman et al. (1958), a leading economics text on the subject, has asserted that Kantorovich, Koopmans and Dantzig were the three joint discoverers of linear programming. But Dorfman's thesis has been disputed by Schwartz (1989) who claimed "that Kantorovich and Koopmans were prediscoverers, and Dantzig alone qualifies as the discoverer." And Gass (1989) endorsed this point of view in his companion piece.

4.5 Non-Uniqueness of the Solution

An unfortunate feature of all linear programming problems including the L_1-norm and L_∞-norm fitting problems is that there may be several solutions to any given problem. that is, there may be several distinct values for the unknown parameters $b_1, b_2, ..., B_p$ corresponding to the same optimal value of the objective function.

Following Hawkins (1993), we shall call any p-dimensional point in parameter space that correspond to a set of p (or more) zero residuals an elemental set solution. Identifying the set of all such solutions, we may obtain their convex hull by forming all nonnegative weighted sums of these elemental set solutions.

In this context, Ben-Tal and Teboulle (1990) have shown that, for any member of a class of strictly isotone functions including the L_1-norm (or sum of absolute values) and the L_2-norm (or sum of squared absolute values) functions, every solution to the problem which chooses values for the parameters $b_1, b_2, ..., b_p$ to minimise the chosen function of the residuals necessarily lies within the convex hull of the set of elemental set solutions. Further, for any member of a class of isotone (but not strictly isotone) functions including the L_∞-norm (or the largest absolute value) and the median (or middlemost) of the (squared) absolute values functions, these authors have shown that at least one of the solutions to the problem which chooses values for the parameters $b_1, b_2, ..., b_p$ to minimise the chosen function of the residuals must lie within the convex hull of the elemental set solutions. It is not necessary for our purposes to know the precise meaning of the terms 'isotone' and 'strictly isotone'; we therefore refer interested readers to the article by Ben-Tal and Teboulle (1990) for definitions.

A possible solution to the problem of non-uniqueness is to consider the family of L_t-norm functions for $1 < t < \infty$. Since each of these functions is strictly convex, the corresponding L_t-norm problems have a unique solution and we may identify the 'true' L_1-norm solution with the value of the (unique) L_t-norm solution in the limit as t tends to unity from above and the 'true' L_∞-norm solution with the value of the (unique) L_t-norm solution in the limit as t tends to infinity from below. Although of considerable theoretical interest, this solution to the problem of non-uniqueness is hardly practical as, in each case, it requires that the *nonlinear L_t-norm* fitting problem be solved for a range of values of t in the vicinity of the desired limiting value.

A second possible solution to the problem of non-uniqueness is to identify the p-dimensional set of solutions to the L_1-norm (or L_∞-norm) problems and to identify the optimal solution with the centre of gravity of this set. This was the solution proposed by Edgeworth (1923, pp. 1076–1077).

As a variation on this theme, Planitz and Gates (1991) have suggested that the optimal solution should be identified with the point in the set of all possible solutions to the L_1-norm (or L_∞-norm) problem which minimizes the sum of the squared residuals. This alternative solution lacks the directness of Edgeworth's approach but it has the advantage that the resulting estimator is unbiased.

And a third possible solution to the problem of non-uniqueness is to replace Edgeworth's deterministic mixture of two or more of the possible solutions with a probabilistic mixture, See Sect. 5.4 below for details.

References

Arthanari, P. S., & Dodge, Y. (1981). *Mathematical Programming in Statistics*. New York: Wiley.

Barrodale, I., & Roberts, F. D. K. (1970). Applications of mathematical programming to L_1 approximation. In J. B. Rosen, O. L. Mangasarian, & K. Ritter (Eds.), *Nonlinear Programming* (pp. 447–464). New York: Academic Press.

Barrodale, I., & Roberts, F. D. K. (1974). Algorithm 478: solution of an overdetermined system of equations in the L_1 norm. *Communications of the Association for Computing Machinery, 17*, 319–320.

Bartels, R., Conn, A. R., & Sinclair, J. W. (1978). Minimization techniques for piecewise differentiable functions: the l_1 solution to an overdetermined linear system. *SIAM Journal of Numerical Analysis, 15*, 224–241.

Ben-Tal, A., & Teboulle, M. (1990). A geometric property of the least squares solution of linear equations. *Linear Algebra and Its Applications, 139*, 165–170. Supplemented (1993) *180*, p. 5.

Bloomfield, P., & Steiger, W. L. (1980), Least absolute deviation curve fiting. *SIAM Journal on Scientfic and Statistical Computing,1*, 290–300.

Bloomfield, P., & Steiger, W. L. (1983). *Least Absolute Deviations*. Boston, Massachusetts: Birkhäuser Publishing.

Bowley, A. L. (1902). Methods of representing the statistics of wages and other groups not fulfilling the normal law of error II: applications to wage statistics and other groups. *Journal of the Royal Statistical Society, 65*, 331–354.

Bowley, A. L., (1928). *F. Y. Edgeworth's Contributions to Mathematical Statistics*. London: The Royal Statistical Society. Reprinted by Augustus, M. Kelley, Clifton, New Jersey, 1972.

Brentjes, S. (1994). Linear optimization. In I. Grattan-Guinness (Ed.), *Companion Encyclopedia to the History and Philosophy of the Mathematical Sciences* (pp. 828–836). London: Routledge.

Cauchy, A. L. (1824). Sur le système des valeurs qu'il faut attribuer à deux éléments déterminés par un grand nombre d'observations pour que la plus grande de toutes les erreurs, abstraction faite du signe, devienne un minimum. *Bulletin de la Soci été Philomatique* (pp. 92–99). Reprinted in his Oeuvres (1958), Series 2, (Vol. 2, pp. 312–322). Paris: Gauthier-Villars.

Cauchy, A. L. (1831). Sur le système de valeurs qu'il faut attribuer à diverses éléments déterminés par un grand nombre d'observations pour que la plus grande de toutes les erreurs, abstraction faite du signe, devienne un minimum. *Journal de l'École Polytechnique, 13*, 175–221. Reprinted in his Oeuvres (1905), Series 2, (Vol. 1, pp. 358–402). Paris, Gauthier-Villars.

Charnes, A., Cooper, W. W., & Ferguson, R. O. (1955). Optimal estimation of executive compensation by linear programming. *Management Science, 1*, 138–151.

Charnes, A., & Cooper, W. W. (1962). On some works of Kantorovich. *Koopmans and others, Management Science, 8*, 246–263.

Dantzig, G. B. (1963). *Linear Programming and Extensions*. Princeton, NJ: Princeton University Press.

de la Vallée Poussin, C. J. (1911). Sur la méthode de l' approximation minimum. *Annales de la Société Scientifique de Bruxelles, 35*, 1–16. English translation by H. E. Salger, National Bureau of Standards, U.S.A.

de Prony, G. C. F. M. (1804). *Recherches physico-mathématiques sur la th éorie des eaux courantes*. Paris: Imprimerie Impériale.

Dodge, Y. (Ed.). (1987). *Statistical data analysis based on the L_1-Norm and related methods*. Amsterdam: North-Holland Publishing Company.

Dodge, Y. (Ed.). (1997). L_1-*Statistical procedures and related topics*. Hayward: Institute of Mathematical Statistics.

Dolby, J. L. (1960). Graphical procedures for fitting the best line to a set of points. *Technometrics, 2*, 477–481.

Dorfman, R. (1984). The discovery of linear programming. *Annals of the History of Computing, 6*, 283–295.

Dorfman, R., Samuelson, P. A., & Solow, R. M. (1958). *Linear Programming and Economic Analysis*. New York: McGraw-Hill.

Edgeworth, F. Y. (1887a). On observations relating to several quantities. *Hermathena, 13*, 279–285.

Edgeworth, F. Y. (1887b). A new method of reducing observations relating to several quantities. *Philosophical Magazine, Series, 5*(24), 222–223.

Edgeworth, F. Y. (1888). On a new method of reducing observations relating to several quantities, Philosophical Magazine, Series 5, 25: 184–191. Reprinted in his Writings, Vol 2. *Edward Elgar, Cheltenham, 1996*, 85–92.

Edgeworth, F. Y. (1923). On the use of medians for reducing observations relating to several quantities. *Philosophical Magazine, Series, 6*(46), 1074–1088.

Farebrother, R. W. (1987). The historical development of the L_1 and L_∞ estimation procedures 1793–1930, in Dodge (pp. 37–63).

Farebrother, R. W. (1988). A simple recursive procedure for the L_1 norm fitting of a straight line. *Applied Statistics, 37*, 457–465.

Farebrother, R. W. (1989). Some early work on the duality between points and lines. *Communications in Statistics (Series B), 18*, 719–727.

Farebrother, R. W. (1992). Least squares initial values for the L_1 norm fitting of a straight line. *Applied Statistics, 41*, 627–633.

Farebrother, R. W. (1997). The historical development of the linear minimax absolute residual estimation procedure 1786–1960. *Computational Statistics and Data Analysis, 24*, 455–466.

Farebrother, R. W. (1999). *Fitting Linear Relationships: A History of the Calculus of Observations 1750–1900*. New York: Springer.

Farebrother, R. W. (2002). *Visualizing Statistical Models and Concepts*. New York: Marcel Dekker. doi:TaylorandFrancis.com.

Farebrother, R. W. (2002b). A dual formulation of Laplace's minimax problem. *Student, 4*, 81–85.

Farebrother, R. W. (2006). A linear programming procedure based on de la Vall ée Poussin's minimax estimation procedure. *Computational Statistics and Data Analysis, 51*, 253–256.

Fisher, W. D. (1961). A note on curve fitting with minimum deviations by linear programming. *Journal of the American Statistical Association, 56*, 359–362.

Fourier, J. B. J. (1827). Second extract from Histoire de l'Académie Royale des Sciences de Paris [pour 1824] (Vol. 7, pp. 47–55). Reprinted in his Oeuvres (1890) (Vol. 2, pp. 325–362). Paris: Gauthier-Villars.

Franksen, O. I. (1985a). Irreversibility by inequality constraints: Part I. On Fourier's inequality. *Systems Analysis in Modelling and Simulation, 2*, 137–149.

Franksen, O. I. (1985b). Irreversibility by inequality constraints: Part III. *Toward mathematical programming, Systems Analysis in Modelling and Simulation, 2*, 337–359.

Gass, S. I. (1989). Comments on the history of linear programming. *Annals of the History of Computing, 11*, 147–151.

Gauss, C. F. (1809). Theoria Motus Corporum Coelestium in Sectionibus Conicis Solem Ambientium. Hampurg: F Perthes and I H Besser. Reprinted in his Werke, (Vol. 7), F. Perthes, Gotha, 1871. English translation by C. H. Davis, Little, Brown and Company, Boston, 1857. Reprinted by Dover, New York, 1963.

Grattan-Guinness, I. (1970). Joseph Fourier's anticipation of linear programming. *Operational Research Quarterly, 21*, 361–364.

Grattan-Guinness, I. (1994). "A new type of question". On the prehistory of linear and nonlinear programming. In E. Knoblock & D. Rowe (Eds.), *The History of Modern Mathematics* (Vol. 3, pp. 43–89). New York: Academic Press.

Harris, T. (1950). Regression using minimum absolute deviations. *The American Statistician, 4*, 14–15.

Hawkins, D. M. (1993). The accuracy of elemental set approximations for regression. *Journal of the American Statistical Association, 88*, 580–589.

Karmarkar, N. (1984). A new polynomial time algorithm for linear programming. *Combinatorica, 4*, 373–395.

Karst, O. J. (1958). Linear curve fitting using least deviations. *Journal of the American Statistical Association*, *53*, 118–132.

Koenker, R. (1997). L_1-computation: An interior monologue, in Dodge (pp. 15–32).

Laplace, P. S. (1786). Mémoire sur la figure de la terre. *Mé moires de l'Académie Royale des Sciences de Paris* [pour 1783], (pp. 17–46). Reprinted in his *Oeuvres Complètes* (1985) (Vol. 1, pp. 3–32). Paris: Gauthier-Villars.

Laplace, P.S. (1793). Sur quelques points du système du monde. *M émoires de l'Académie Royale des Sciences de Paris* [pour 1789], (pp. 1–87). Reprinted in his *Oeuvres Complètes* (1985), (Vol. 11, pp. 447–558). Paris: Gauthier-Villars.

P. S. Laplace (1799), Traité de Mécanique Céleste, Tome II, J. B. M. Duprat, Paris. Reprinted in his *Oeuvres Complètes* (1843), (Vol. 2), Imprimerie Royale, Paris, 1843 and Gauthier-Villars, Paris, 1878. English translation by N. Bowditch, Hillard, Gray, Little and Wilkins, Boston, 1832. Reprinted by Chelsea Publishing Company, New York, 1966.

Laplace, P. S. (1812), Théorie Analytique des Probabilités. Paris: Courcier. Third edition with an introduction and three supplements, Courcier, Paris, 1820. Reprinted in his Oeuvres Compl ètes, (Vol. 7), Imprimerie Royale, Paris, 1847 and Gauthier-Villars, Paris, 1886.

Laplace, P. S. (1818). Deuxième Supplément to Laplace (1812).

Owen, D. B., & Shiau, J. J. H. (1988). On the duality between points and lines. *Communications in Statistics (Series A)*, *17*, 207–228.

Portnoy, S. (1997). On computation of regression quantiles: Making the Laplacian tortoise faster, in Y. Dodge (Ed.) (pp. 187–200).

Planitz, M., & Gates, J. (1991). Strict discrete approximation of the L_1 and L_∞-norms. *Applied Statistics*, *40*, 113–122.

Portnoy, S., & Koenker, R. (1997). The Gaussian hare and the Laplacian tortoise: Computability of squared-error v. absolute-error estimators. *Statistical Science*, *12*, 279–300.

Rhodes, E. C. (1930). Reducing observations by the method of minimum deviations. *Philosophical Magazine, Series*, *7*(9), 974–992.

Schwartz, R. (1989). The invention of linear programming. *Annals of the History of Computing*, *11*, 145–147.

Singleton, R. R. (1940). A method for minimising the sum of absolute values of deviations. *Annals of Mathematical Statistics*, *11*, 301–310.

Souvaine, D. L., & Steele, J. M. (1987). Time and space-efficient algorithms for least median of squares regression. *Journal of the American Statistical Association*, *82*, 794–801.

Stay, B. (1760). Philosophiae Recentioris... Versibus Traditae..., Tomus II, annotated by R. J. Boscovich, N. and M. Palearini, Rome. French translation of pp. 385–397 included in the French translation of Maire and Boscovich (1755, pp. 501–506).

Stiefel, E. (1959). Uber diskrete und lineare Tschebyscheff - Approximationen. *Numerische Mathematik*, *1*, 1–28.

Stiefel, E. (1960). Note on Jordan elimination, linear programming and Tchebyscheff approximation. *Numerische Mathematik*, *2*, 1–17.

Turner, H. H. (1887). On Mr. Edgeworth's method of reducing observations relating to several quantities. *Philosophical Magazine*, *24*, 466–470. Series 5.

Wagner, H. M. (1959). Linear programming techniques for regression analysis. *Journal of the American Statistical Association*, *54*, 202–212.

Watson, G. A. (2000). Approximation in normed linear spaces. *Journal of Computational and Applied Mathematics*, *121*, 1–36.

Williams, H. P. (1986). Fourier's method of linear programming and its dual. *American Mathematical Monthly*, *93*, 681–695.

Chapter 5
Statistical Theory

Abstract Supposing that the disturbance terms in the standard linear statistical model are independent and follow a common Laplacian, Gaussian, or rectangular distribution, then the principle of maximum likelihood suggests that we should choose estimates of the slope parameters to minimise the L_t-norm of the residuals with $t = 1$, $t = 2$ or $t = \infty$ respectively. In this context, we outline the small sample and asymptotic theory relating to these maximum likelihood estimators and the related Likelihood Ratio, Lagrange Multiplier and Wald tests of linear restrictions on the parameters of the model. We also demonstrate that a simple modification of the standard linear programming implementation of the l_1-norm or L_∞-norm fitting problem yields (pseudo-unbiased) estimators that are symmetrically distributed about the true parameter values when the disturbances are symmetrically distributed about zero.

Keywords Gaussian disturbances · Gauss-Markov linear model · Huber's M-estimation · L_p-norm estimation · Laplacian disturbances · Linear statistical model · Maximum likelihood estimation · Pseudo-Unbiased estimation · Uniform disturbances.

5.1 Linear Statistical Model

Having progressed thus far in this brief without the aid of a formal statistical model, we now generalise the algebraic scheme of Chap. 3 by supposing that, For $i = 1, 2, ..., n$, the set $\{x_{i1}, x_{i2}, ..., x_{ip}, y_i\}$ represents the ith observation on a set of $p+1$ variables $\{X_1, X_2, ..., X_p, Y\}$, and suppose that we wish to fit a linear statistical model of the form

$$y_i = x_{i1}\beta_1 + x_{i2}\beta_2 + \cdots + x_{ip}\beta_p + \epsilon_i$$

R. W. Farebrother, L_1-*Norm and* L_∞-*Norm Estimation,*
SpringerBriefs in Statistics, DOI: 10.1007/978-3-642-36300-9_5,
© The Author(s) 2013

to these n observations, where $\beta_1, \beta_2, ..., \beta_p$ are a set of p fixed but unknown parameters and $\epsilon_1, \epsilon_2, ..., \epsilon_n$ are a set of n stochastic disturbance terms whose statistical properties will be discussed below.

Expressing this model in matrix form, we have

$$y = X\beta + \epsilon$$

where y, X, β and ϵ are $n \times 1$, $n \times p$, $p \times 1$ and $n \times 1$ matrices respectively.

5.2 Maximum Likelihood Estimation

In the context of the model of Sect. 5.1 with n independent disturbance terms each of which is known (or supposed) to follow a common law of distribution, most readers will know that the Principle of Maximum Likelihood often suggests a suitable estimation procedure.

In the most familiar case, if we know (or suppose) that the disturbance terms are independently and identically distributed with a Gaussian or normal distribution with a zero mean and a finite variance, then the maximum likelihood principle suggests that we should choose an (L_2-norm) estimator $\tilde{\beta}(2)$ for β in such a way as to minimise the sum of the squared estimated disturbances (or residuals)

$$\sum_{i=1}^{n} e_i^2.$$

Similarly, if we know (or suppose) that the disturbance terms are independently and identically distributed with a Laplacian or double exponential distribution with a zero mean and a finite variance, then the maximum likelihood principle suggests that we should choose an (L_1 -norm) estimator $\tilde{\beta}(1)$ for β in such a way as to minimise the sum of the absolute estimated disturbances (or residuals)

$$\sum_{i=1}^{n} |e_i|.$$

Finally, if we know (or suppose) that the disturbance terms are independently and identically distributed with a uniform or rectangular distribution with a zero mean and a finite range, then the maximum likelihood principle suggests that we should choose an (L_∞-norm) estimator $\tilde{\beta}(\infty)$ for β in such a way as to minimise the largest absolute estimated disturbance term (or residual)

$$max_{i=1}^{n} |e_i|.$$

Thus, in the context of maximum likelihood estimation, the L_2-norm and L_1-norm fitting procedures are intimately associated with the Gaussian (normal) and the Laplacian (double exponential) distributions respectively. In addition, Gauss had developed much of the finite sample theory relating to the L_2-norm estimator $\hat{\beta} = \tilde{\beta}(2)$ whilst Laplace had performed a similar service in respect of the asymptotic theory for the L_1-norm estimator $\tilde{\beta}(1)$. and, for this reason, Portnoy (1997) and Portnoy and Koenker (1997) have suggested that we should associate the L_2-norm fitting procedure with the name of Gauss and the L_1-norm procedure with that of Laplace.

Further, with conventional computational techniques, the L_2-norm fitting procedure is usually much faster than the L_1-norm procedure, and these authors again thought it natural to associate these procedures with the hare and the tortoise from Aesop's fable, whence the exotic element in the titles of the articles by Portnoy (1997), Portnoy and Koenker (1997) and Farebrother (2002b).

Unfortunately, Portnoy and Koenker's statistical bestiary breaks down when we turn to the L_∞-norm fitting procedure. Although anyone wishing to extend this scheme to the L_∞-norm procedure will have little difficulty in identifying the turtle as the '*dual*' of a tortoise, it is not possible to suggest a single name to associate with both the uniform (or rectangular) distribution and the L_∞-norm fitting procedure. On the other hand, it seems sensible to make a virtue of necessity by following the practice of approximation theory and associating the name of Chebyshev with the L_∞-norm fitting procedure if not with that of the uniform distribution.

5.3 Asymptotic Theory

It is well-known that the unweighted L_t-norm estimator $\tilde{\beta}(t)$ of β with $w_A = w_B$ equal to a column of ones is a member of Huber's class of M-estimators. For sufficiently small values of $t > 1$, we may therefore deduce that the function $\sqrt{n}[\tilde{\beta}(t)-\beta]$ is asymptotically normally distributed with mean 0 and variance $\sigma_t^2 \Omega^{-1}$ where

$$\Omega = Lim_{n \to \infty}(\frac{1}{n}X'X)$$

is a $p \times p$ positive definite matrix and where σ_t^2 is a positive scalar. Applying this general result in the limit as t tends to unity from above we find that the function $\sqrt{n}[\tilde{\beta}(1) - \beta]$ is asymptotically normally distributed with mean 0 and variance $\sigma_1^2 \Omega^{-1}$ where $\tilde{\beta}(1)$ is the unique L_1-norm estimator of β defined in Sect. 4.5 and σ_1^2 is the asymptotic variance of the sample median from random samples of n disturbance terms generated by the model of Sect. 5.1 (as may readily be established by setting $p = 1$ and X equal to an $n \times 1$ column of ones).

In this connection, Koenker and Bassett (1982) have shown that the L_1-norm equivalents of the familiar Lagrange Multiplier, Likelihood Ratio and Wald tests of a set of linear constraints on the β parameters of the linear statistical model of

Sect. 5.1 have an asymptotic chi-squared distribution. For further details, see Bassett and Koenker (1978), Koenker and Bassett (1985) and Koenker (1987).

5.4 Pseudo-Unbiased Weighted L_1-Norm Procedures

If we restrict ourselves to the class of symmetrically weighted L_t-norm estimators of β, that is, to the case in which $w_A = w_B = w$ is an $n \times 1$ matrix with nonnegative elements, possibly an $n \times 1$ column of ones, define $\hat{\beta} = c^* = (X'X)^{-1}X'y$ and subtract $X\hat{\beta}$ from both sides of the equality constraint in Problem P_1 of Sect. 3.4, set $y^* = y - X\hat{\beta} = My$ and $a = a_A - a_B = b - \hat{\beta}$ where a_A and a_B are $p \times 1$ nonnegative matrices, then our first variant of Problem P_1 may be written as

Problem $P_1^{(1)}$:
Minimise $w'e_A + w'e_B$
subject to
$$X a_A - X a_B + e_A - e_B = y^*$$
and $a_A \geq 0,\ a_B \geq 0,\ e_A \geq 0,\ e_B \geq 0.$

Interchanging the roles of the $p \times 1$ matrices a_A and a_B at the same time as interchanging the roles of the $n \times 1$ matrices e_A and e_B, we have a second (equivalent) variant of Problem P_1:

Problem $P_1^{(2)}$:
Minimise $w'e_B + w'e_A$
subject to
$$X a_B - X a_A + e_B - e_A = -y^*$$
and $a_B \geq 0,\ a_A \geq 0,\ e_B \geq 0,\ e_A \geq 0.$

Replacing the $n \times 1$ matrix of disturbance terms ϵ by its negation $-\epsilon$ in the above problems, we find that the formulation of Problem $P_1^{(1)}$ is the same as that of Problem $P_1^{(2)}$ except that the roles of a_A and a_B have been reversed as have those of e_A and e_B. In this context, we may readily establish that the L_1-norm estimator of β implied by the solution of the joint problem is symmetrically distributed about β provided that these two variants of Problem P_1 are each chosen with probability 0.5 and provided that, for each choice of an $n \times 1$ matrix v, the $n \times 1$ matrix of disturbances ϵ is equally likely to take the values $\epsilon = +v$ and $\epsilon = -v$.

In this context, the $n \times 1$ matrices ϵ and $y^* = My = M\epsilon$ are both symmetrically distributed about the $n \times 1$ matrix 0, the $p \times 1$ matrix $a = a_A - a_B$ is symmetrically distributed about the $p \times 1$ matrix 0, and the $p \times 1$ matrix $b = a + \hat{\beta}$ is symmetrically distributed about β. Thus, the L_1-norm estimator $b = a + \hat{\beta}$ is necessarily an unbiased estimator of β provided that its mean exists.

For more general results on the unbiasedness of the L_1-norm estimator, see Sielken and Hartley (1973), Harvey (1978) and Sposito (1982).

5.5 Pseudo-Unbiased Weighted L_∞-Norm Procedures

Contrary to expectations, the technique employed in 5.4 is not restricted to L_t-norm procedures with t in the range $0 < t \leq 2$. To see this, we first restore the traditional notation to Problem P_5 of Sect. 3.3 before rewriting it as

Problem $P_5^{\#}$:
Minimise k
subject to
$$Xb + kw_B \geq y$$
$$-Xb + kw_A \geq -y.$$

Again, setting $w_A = w_B = w$ and subtracting $X\hat{\beta}$ from both sides of both inequality constraints in Problem $P_5^{\#}$, then our first variant of this Problem may be written as

Problem $P_5^{(1)}$:
Minimise k
subject to
$$Xa_A - Xa_B + kw \geq y^*$$
$$-Xa_A + Xa_B + kw \geq -y^*$$
and $a_A \geq 0, a_B \geq 0.$

Again, interchanging the roles of the $p \times 1$ matrices a_A and a_B at the same time as interchanging the roles of the $n \times 1$ matrices e_A and e_B, we have our second (equivalent) variant of Problem $P_5^{\#}$:

Problem $P_5^{(2)}$:
Minimise k
subject to
$$Xa_B - Xa_A + kw \geq -y^*$$
$$-Xa_B + Xa_A + kw \geq y^*$$
and $a_B \geq 0, a_A \geq 0.$

Again, as in Sect. 5.4, we may readily show that the L_∞-norm estimator of β is symmetrically distributed about β when ϵ is symmetrically distributed about 0 provided that we choos Problems $P_5^{(1)}$ and $P_5^{(2)}$ with probability 0.5.

Finally, it is of interest to note that the simple idea outlined in Sects. 5.4 and 5.5 may also be extended to the case in which w_A is a scale multiple of w_B, and, in particular, to the case in which w_A and w_B are both positive multiples of the $n \times 1$ column of ones, that is, to the family of regression quantiles defined in Sect. 3.3, see Farebrother (1985a) for details.

References

Bassett, G. W., & Koenker, R. (1978). The asymptotic theory of the least absolute error estimator. *Journal of the American Statistical Association, 73*, 618–622.

Dodge, Y. (Ed.). (1987). *Statistical Data Analysis Based on the L_1-Norm and Related Methods.* Amsterdam: North-Holland Publishing Company.

Dodge, Y. (Ed.). (1997). *L_1-Statistical Procedures and Related Topics.* Hayward: Institute of Mathematical Statistics.

Dodge, Y. (Ed.). (2002). *Statistical Data Analysis based on the L_1-Norm and Related Methods.* Basel: Birkhäuser Publishing.

Farebrother, R. W. (2002b). Whose hare and whose tortoise, in Y. Dodge (Ed.) (pp. 253–256).

Farebrother, R. W. (1985a). Unbiased L_1 and L_∞ estimation. *Communications in Statistics (Series A), 14*, 1941–1962.

Harvey, A. C. (1978). On the unbiasedness of robust regression estimators. *Communications in Statistics (Series A), 7*, 779–783.

Koenker, R. (1987). A comparison of asymptotic testing methods for L_1 regression, in Y. Dodge (Ed.) (287–296).

Koenker, R., & Bassett, G. W. (1982). Tests of Linear Hypotheses and L_1 Estimation. *Econometrica, 50*, 1577–1583.

Koenker, R., & Bassett, G. W. (1985). On Boscovich's estimator. *Annals of Statistics, 13*, 1625–1628.

Portnoy, S. (1997). On computation of regression quantiles: Making the Laplacian tortoise faster, in Y. Dodge (Ed.) (pp. 187–200).

Portnoy, S., & Koenker, R. (1997). The Gaussian hare and the Laplacian tortoise: Computability of squared-error v. absolute-error estimators. *Statistical Science, 12*, 279–300.

Sielken, R. L., & Hartley, H. O. (1973). Two linear programming algorithms for unbiased estimation of linear models. *Journal of the American Statistical Association, 68*, 639–641.

Sposito, V. A. (1982). On unbiased L_1 regression estimators. *Journal of the American Statistical Association, 77*, 652–653.

Chapter 6
The Least Median of Squared Residuals Procedure

Abstract The formal results outlined in Chap. 5 cease to be valid if the disturbance terms depart from a strict adherence to the assumptions underlying the statistical model of Chap. 5. In the present chapter we seek to extend the robustness properties of the L_1-norm estimator to a variant of the L_∞-norm estimator by ignoring about one-half of the available data and applying the L_∞-norm procedure to the remaining observations. The estimator defined in this way is known as the least median of squares (LMS) estimator and has excellent statistical properties provided that no more than one-half of the observations are outliers. Unfortunately, it can be extremely expensive to determine the true LMS estimates and practitioners often resort to techniques which identify a close approximation to these estimates from a random sample of elemental set (or subset) estimates.

Keywords Elemental set estimators · Jump discontinuity · Least median of squared residuals · Minimum volume ellipsoid · Nonlinear programming · Robustness to outliers · Subset estimators

6.1 Robustness to Outliers

Although the fitting techniques described in Chap. 5 are of some theoretical interest, it does not seem reasonable to assume that the disturbance terms will exactly satisfy any specific set of assumptions and thus our principal justification for using variants of the L_1-norm and L_∞-norm fitting procedures must be on the basis of their relative robustness to variations in the distributional assumptions underlying the statistical model outlined in Sect. 5.1. This is an immense topic which we do not propose to discuss here. Instead, the interested reader is referred to Huber (1981, 1987), Hampel (1974) and Hampel (1986) for details.

R. W. Farebrother, *L₁-Norm and L∞-Norm Estimation*,
SpringerBriefs in Statistics, DOI: 10.1007/978-3-642-36300-9_6,
© The Author(s) 2013

6.2 Gauss's Optimality Conditions

Thus far, the L_1-norm fitting procedure seems to has been carrying the alternative L_∞-norm procedure along in the guise of a poor relation, but now a variant of the latter method comes into its own as a procedure that is more robust to the presence of outlying observations.

As noted in Sect. 4.1, Gauss (1809) established a set of optimality conditions for the unweighted L_1-norm procedure (whether constrained to pass through the centroid or not). However, he seems to have been worried that his optimality conditions implied that, after the full set of n observations had determined which p of them should define the optimal L_1-norm estimator, the remaining $n - p$ observations did not participate further in the actual determination of values for the p unknown parameters. Indeed, he noted that the y values of the $n - p$ omitted observations could be varied to any extent without affecting the result so long as any such variation preserved the positive or negative signs attached to the corresponding residuals.

Conversely, Appa and Smith (1973) have shown that the L_∞-norm fitting procedure is often extremely sensitive to the displacement of one (or more) of the y-observations. That is, it is extremely non-robust to the presence of one or more outlying observations.

To counteract the problem of outlying observations of this type, Rousseeuw (1984) has suggested implicitly that we should transfer the property of extreme robustness to certain types of outlying observations demonstrated by the L_1-norm procedure to the L_∞-norm procedure by requiring that the resulting estimator should ignore a large proportion (usually one-half) of the observations. Rousseeuw's implementation of this general idea defines the so-called Least Median of Squares (or LMS) fitting procedure which chooses values for the parameters b_1, b_2, ..., b_p to minimise the median or middlemost of any increasing function of the absolute residuals; and, in particular, it chooses values for the parameters to minimise the median or middlemost of the squared (absolute) residuals, $Med(e_i^2)$, where the median or middlemost of a set of n observations is formally defined in Sect. 2.1.

6.3 Least Median of Squares Computations

Conditional on the choice of the particular set of m observations to be retained, we have seen that the LMS problem takes the form of a L_∞-norm fitting problem which may be expressed as a linear programming problem. However, on removing the assumption that we know which m observations are to be retained, we find that the full LMS procedure takes the form of a mixed integer linear programming problem in which m of the n residuals are associated with unit indicators whilst the remaining $n - m$ residuals have zero indicators.

Thus, if we explicitly incorporate a set of n indicators z_1, z_2, ..., z_n, the ith of which takes a unit value if the ith residual features in the calculations and a zero value if not, then our problem becomes one of choosing values for the parameters $b_1, b_2, \ldots, b_p$ to solve the mixed integer linear programming problem:

Problem LMS:

Minimise $\sim k$

subject to

$$-k \leq e_i \leq k \quad \text{when } z_i = 1$$

$$\sum_{i=1}^{n} z_i = m$$

and $\quad z_i = 0 \text{ or } z_i = 1$

where, for $i = 1, 2, \ldots, n$, the ith residual is defined by

$$e_i = y_i - x_{i1}b_1 - x_{i2}b_2 - \cdots - x_{ip}b_p.$$

However, the sheer complexity of the computations associated with a formal implementation of the LMS fitting problem implies that the computational procedures are likely to be impractical. In this context, we are therefore often obliged to substitute an approximate procedure for the full formal LMS procedure. In particular, we might seek to approximate the optimal values of the p parameters by evaluating the median squared residual function $Med(e_i^2)$ for a sufficiently large sample of the L_∞-norm determinations of the p unknowns from a random sample of arbitrarily chosen sets of $p+1$ equations. However, it is clear that the L_∞-norm fitting of a system of $p+1$ equations in p unknowns is vastly more expensive than the direct solution of a set of p equations in p unknowns. Rousseeuw and Leroy (1987) have therefore suggested that a sufficiently accurate approximation to the exact LMS solution may be obtained by evaluating the median squared residual function $Med(e_i^2)$ for a sufficiently large sample of (Gaussian) elemental set determinations corresponding to sets of p zero residuals. This conjecture was subsequently confirmed by Stromberg (1993) whilst Hawkins (1993) has shown that this approximation technique yields satisfactory results for a wide class of robust fitting procedures. [Note, in passing, that this class of elemental set estimators are also employed in the detection of outlying observations, see Farebrother (1988, 1997) and Hawkins et al. (1984).]

6.4 Minimum Volume Ellipsoid

The general Least Median of Squares hyperplane-fitting procedure of Sects. 6.2 and 6.3 may be specialised to yield the corresponding point-fitting procedure. Given a set of n points in p-dimensional space and a suitable value for m close to $n/2$, we may consider the set of all p-dimensional spheres which just contain m of these n points, select the sphere of minimal p-dimensional volume (and hence minimal radius), and identify the centre of this optimal sphere as our estimate of the centre of the distribution.

As a minor generalisation of this procedure, we may consider the set of all p-dimensional ellipsoids which just contain m of the n given points, select the ellipsoid of minimal p-dimensional volume, and again identify the centre of this optimal ellipsoid as our estimate of the centre of the distribution. Rousseeuw (1984) refers to the latter estimator as the 'minimum volume ellipsoid'. For a further generalisation to minimum volume cylinders, see Farebrother (1992, 1994).

In the one-dimensional case, any ellipsoid or sphere will degenerate to a one-dimensional line segment and, as in Sect. 2.1, our general procedure defines the midpoint of the shortest line segment which just covers m of the n points. If m is close to $n/2$, then, as noted in Sect. 2.1, this midpoint is known as the 'shortest half'.

By definition, any of the estimators described in this chapter are extremely robust to the presence of outlying observations as almost one-half of the observations can take arbitrarily large (positive or negative) values without much affecting the value returned by the relevant estimation procedure. They are thus likely to command significant roles in the area of robust statistical analysis.

On the other hand, Hettmansperger and Sheather (1992) have shown that the fitted values returned by these procedures can exhibit jump discontinuities in response to smooth variations in the data. But this result was already known in principle some seventy years earlier as Whittaker and Robinson (1924, 1944, pp. 215–217) have shown that the arithmetic mean is the only twice differentiable function of the observations on a single variable which is equivariant to changes of location, equivariant to changes of scale, and invariant to permutations of the order of the observations, see Farebrother (1999, p. 80). Moreover, this result can be combined with Farebrother's (1989) algebraic characterisation of the least squares estimator to yield a generalisation that applies to the linear model of Sect. 5.1 with two or more explanatory variables.

References

Appa, G., & Smith, C. (1973). On L_1 and Chebychev estimation. *Mathematical Programming, 5,* 73–87.

Dodge, Y. (Ed.). (1987). *Statistical Data Analysis Based on the L_1-Norm and Related Methods.* Amsterdam: North-Holland Publishing Company.

Dodge, Y. (Ed.). (1992). *L_1-Statistical Analysis and Related Methods.* Amsterdam: North-Holland Publishing Company.

Dodge, Y. (Ed.). (1997). *L_1-Statistical Procedures and Related Topics.* Hayward: Institute of Mathematical Statistics.

Farebrother, R. W. (1988). Elemental location shift estimators in the constrained linear model. *Communications in Statistics (Series A), 17,* 79–85.

Farebrother, R. W. (1989). Systems of axioms for estimators of the parameters of the standard linear model. *Oxford Bulletin of Economics and Statistics, 51,* 91–94.

Farebrother, R. W. (1992). The geometrical foundations of a class of estimation procedures which minimise sums of Euclidean distances and related quantities, in Dodge (pp. 337–349).

Farebrother, R. W. (1994). On hyperplanes of closest fit. *Computational Statistics and Data Analysis, 19,* 53–58.

Farebrother, R. W. (1997). Notes on the early history of elemental set methods, in Dodge (pp. 161–170).

Farebrother, R. W. (1999). *Fitting Linear Relationships: A History of the Calculus of Observations 1750–1900*. New York: Springer.

Hampel, F. R., Ronchetti, E. M., Rousseeuw, P. J., & Stahel, W. A. (1986). *Robust Statistics : The Approach Based on Influence Functions*. New York: Wiley

Gauss, C. F. (1809). Theoria Motus Corporum Coelestium in Sectionibus Conicis Solem Ambientium. Hamburg, F Perthes and I H Besser. Reprinted in his Werke, (Vol. 7) F. Perthes, Gotha, 1871. English translation by C. H. Davis, Little, Brown and Company, Boston, 1857. Reprinted by Dover, New York, 1963.

Hampel, F. R. (1974). The influence curve and its role in robust estimation. *Journal of the American Statistical Association, 69*, 383–393.

Hawkins, D. M. (1993). The accuracy of elemental set approximations for regression. *Journal of the American Statistical Association, 88*, 580–589.

Hawkins, D. M., Bradu, D., & Kass, C. V. (1984). Location of several outliers in multiple regression data using elemental sets. *Technometrics, 19*, 197–208.

Hettmansperger, T. P., & Sheather, S. J. (1992). A cautionary note on the method of least median of squares. *The American Statistician, 46*, 79–83.

Huber, P. J. (1981). *Robust Statistics*. New York: Wiley.

Huber, P. J. (1987). The place of the L_1-norm in robust estimation. *Computational Statistics and Data Analysis 5*, 255–262. Reprinted in Dodge (pp. 23–33).

Rousseeuw, P. J. (1984). Least median of squares regression. *Journal of the American Statistical Association, 79*, 871–880.

Rousseeuw, P. J., & Leroy, A. M. (1987). *Robust Regressions and Outlier Detection*. New York: Wiley.

Stromberg, A. J. (1993). Computing the exact value of the least median of squares estimate and stability diagnostics in multiple linear regression. *SIAM Journal of Scientific Computing, 14*, 1289–1299.

Whittaker, E. T., & Robinson, G. (1924). *The Calculus of Observations*. London: Blackie. Fourth edition (1944) reprinted by Dover, New York, 1967.

Chapter 7
Mechanical Representations

Abstract We describe a set of mechanical models that may be used to represent the various L_1-norm, L_2-norm and L_∞-norm fitting procedures: the L_1-norm estimation problems may be represented by the positioning of a ring or a rigid rod under the influence of a frictionless system of strings and pulleys; the L_2-norm estimation problems may be represented by the positioning of a ring or a rigid rod under the influence of a frictionless system of stretched springs; and, by combining disparate aspects from these two mechanical models, we find that L_∞-norm estimation problems may be represented by the positioning of a ring or a rigid rod under the influence of a system of strings and blocks. In the first two cases the optimal position of the ring or rigid rod is determined by a minimisation of the total potential energy of the system. In the third case we only have to determine the physical limitations imposed by the lengths of string attached to the ring or rod. Moreover, the mechanical model for the L_1-norm problem may be generalised to cover Oja's bivariate median and the L_∞-norm model may be generalised to cover Rousseeuw's least median of squares problem.

Keywords Constrained and unconstrained estimation · Fermat's facility location problem · Graph theory · Influential observations · Mechanical models based on springs · Mechanical models based on strings and blocks · Mechanical models based on strings and pulleys · Oja's bivariate median · Potential energy · Rogerius Josephus Boscovich (1711–1787) · William Fishburn Donkin (1814–1869) · Pierre de Fermat (1601–1665).

7.1 Introduction

As in Chap. 2 and Sect. 3.1, we restrict ourselves to the special case in which we suppose that we have been given a set of n matched pairs of observations on two variables X and Y. Then, as noted in Sect. 4.1, these n observations may either be

R. W. Farebrother, L_1-Norm and L_∞-Norm Estimation,
SpringerBriefs in Statistics, DOI: 10.1007/978-3-642-36300-9_7,

represented as n points in the xy-plane of observations, or as n straight lines in the ba-plane of parameters. Further, we may be interested in fitting a single point or a straight line to these observations. This yields a total of four distinct problems of interest. Most of the present Chapter is concerned with outlining L_1-norm methods for fitting a point or a line to the set of points in the xy-plane of observations and a point to this set of lines in the ba-plane of parameters. It is also possible to conceive of fitting a line in the second case, but this problem has not yet been addressed in the literature and we do not propose to initiate its discussion here. Instead, we shall take the opportunity of discussing a variant of our third problem adapted for use with Oja's (1983) bivariate median.

Several authors have suggested that mechanical models based on strings and pulleys can be constructed for the first of these fitting problems. In this chapter we will show that this mechanical model can easily be extended to all three variants of the basic L_1-norm fitting problem mentioned above.

7.2 Fitting a Point to a Set of Points in the Plane of Observations: Fermat's Problem

For our first problem, we suppose that the n pairs of observations on the variables X and Y are represented as n points (x_i, y_i) $(i = 1, 2, ..., n)$ in the xy-plane of observations. Defining an additional point (x_0, y_0) in the same plane, we seek to determine the values of x_0 and y_0 which minimize the weighted sum of the absolute Euclidean distances

$$\sum_{i=1}^{n} w_i \sqrt{[(x_i - x_0)^2 + (y_i - y_0)^2]}$$

between this additional point and the n given points where w_1, w_2, ..., w_n are a set of known positive weights.

As a variant of this unconstrained problem, we may suppose that the additional point (x_0, y_0) is constrained to lie on the straight line $y = a_0 + b_0 x$ by imposing the condition $y_0 = a_0 + b_0 x_0$ for some suitable values of a_0 and b_0. Indeed, if appropriate, we may suppose that the additional point is constrained to lie on a curved line or to lie in a region of the xy-plane, see Farebrother (2002) for details.

The unconstrained problem was first posed by Fermat in 1638 in the case when $n = 3$ and $w_1 = w_2 = w_3$. However, the absence of distinct weights from the original statement of the unconstrained problem is probably not significant as in August 1657 and again in May 1662 Fermat proposed two constrained variants of the problem relating to optical refraction in which $n = 2$ and $w_2 = 2w_1$, see Farebrother (1990) for details. [By contrast, Kuhn (1967, 1973) has suggested that the first weighted generalisation of Fermat's problem is due to Simpson in 1750].

The special case of the unconstrained and unweighted problem in which the number of observations exceeds the number of dimensions by unity has an interesting history culminating in the first dual nonlinear programming problem posed and solved by Fasbender in 1846. The history of this problem has been ably summarised by Kuhn (1967) . However, as a supplement to this account, it should be noted that, for $m = 2, 3, ..., n$, the m vectors pointing from the centre of a regular $(m - 1)$-dimensional simplex to its m vertices (which characterise the geometrical solution of this problem) are intimately associated with the $m - 1$ vectors defined by Helmert's transformation, and thus with the associated set of recursive residuals which are employed in the statistical detection of departures from the assumptions underlying the linear statistical model of Sect. 5.1, see Farebrother (1985, 1988a, 1999, 2002) for further details.

The abstract mathematical problem outlined above may be given a physical form by associating a real horizontal plane with the Cartesian xy-plane of observations. Given such a plane, we drill holes through it at the points (x_i, y_i) corresponding to the n observations on X and Y. We attach weighted strings running from the given points to a ring at the arbitrary point (x_0, y_0). Now, the ring may be supposed to have moved a distance

$$\sqrt{[(x_i - x_0)^2 + (y_i - y_0)^2]}$$

from the ith hole at (x_i, y_i) to the arbitrary point (x_0, y_0), and consequently the ith weight may be considered to have moved vertically through the same distance and thus has gained potential energy proportional to

$$w_i \sqrt{[(x_i - x_0)^2 + (y_i - y_0)^2]}.$$

So that, the system as a whole has potential energy proportional to

$$\sum_{i=1}^{n} w_i \sqrt{[(x_i - x_0)^2 + (y_i - y_0)^2]}.$$

When the ring is not at the ith hole, the ith weight will induce a force proportional to w_i tending to pull the ring from its present position towards the ith hole. And, when the system is in a state of equilibrium, the position of the ring will identify the values of x_0 and y_0 that define the mediancentre whose unweighted variant is defined in Sect. 2.2.

By contrast with the familiar L_2-norm point fitting problem outlined in Sect. 7.6 below, the present model has little to offer besides the observation that when joined head to tail the sequence of line segments representing the n forces acting on the ring must form a closed circuit. Thus, in the special case when all the observations lie on a straight line, this procedure defines the weighted median whose unweighted variant is defined in Sect. 2.1. Further, if there are just three observations in the plane with equal weights, then, selecting one of these unit forces as our prime direction and resolving the three unit forces parallel to this direction and at right angles to it, reveals that this set of vectors necessarily point to the three vertices of an equilateral triangle (or regular two-dimensional simplex) which, as mentioned above, characterises the geometrical solution of this problem.

A variant of this mechanical model (which took greater care to eliminate friction by substituting pulleys for the holes drilled through the horizontal board) was first developed by Varignon in 1687 and employed by Lamé and Clapeyron in 1829. The model is also associated with the name of Alfred Weber (brother of Max) although the section of his book of 1909 in which this model was developed was actually written by his colleague Georg Pick who apparently borrowed the model from an 1883 work of the physicist Ernst Mach, see Franksen and Grattan-Guinness (1989) for a detailed discussion of the history of this problem and an English translation of Lamé and Clapeyron's article.

In passing, we note that Lamé and Clapeyron were also concerned with the modelling of non-Euclidean distances in the xy-plane; thus giving rise to a mechanical representation of graph theoretical problems and the concept of the median of a graph. Farebrother (2002) has shown that this model may easily be generalised to higher dimensions by replacing the strings confined to certain specified paths in the plane by strings in flexible hollow tubes in q-dimensional space.

Lamé and Clapeyron were also concerned with the effect of varying the weights attached to the strings. Thus, in their military example, they found that the supply depot of an army corps should not necessarily be placed at the same point as its headquarters as the former should be weighted by the quantities of goods required by the various constituent units and the latter by the number of staff officers attached to these units. Further, they found that the movement of one or more army units would affect the optimal location of these two facilities in different ways.

7.3 Fitting a Line to a Set of Points in the Plane of Observations: Boscovich's Problem

For our second problem, as in Sect. 3.1, we suppose that the n pairs of observations on the variables X and Y are again represented as n points $(x_i,\ y_i)$ $(i = 1, 2, ..., n)$ in the xy-plane of observations. Defining an arbitrary line $ya+bx$ in the same plane, we seek to determine the values of the parameters a and b which minimize the weighted sum of the absolute x-meridian distances from the n points to the arbitrary line

$$\sum_{i=1}^{n} w_i |y_i - a - bx_i|$$

where these x-meridian distances are measured parallel to the y-axis.

Again, we may define a constrained variant of this second problem by supposing that the arbitrary line must pass through the point (x_0, y_0) by imposing the condition $a = y_0 - bx_0$. For example, if we set $x_0 = \sum x_i/n$ and $y_0 = \sum y_i/n$ in the unweighted case, then this condition becomes the familiar adding-up constraint

$$\sum_{i=1}^{n}(y_i - a - bx_i) = 0$$

employed in the constrained variant of this problem proposed by Boscovich in a report to the Academy of Bologna in 1757 and again in his notes to a poem in Latin hexameters by Stay in 1760. [In this context, it is pertinent to note that these dates refer to a period some fifty years before Legendre first published his account of the method of least squares in 1805].

This constrained variant of the problem was also briefly examined by Simpson in June 1760 and, at far greater length, by Laplace (1793, 1799, 1812, 1818), see Eisenhart (1961), Farebrother (1990, 1993, 1999) and Stigler (1984, 1986) for details.

As in Sect. 7.2, we have to associate a real horizontal plane with the abstract Cartesian plane of observations, then at each point (x_i, y_i) associated with an observation, we have to drill a hole. Through this hole we pass a length of string with a weight w_i at the lower end and a small ring at the upper end. We now place a rigid rod at an arbitrary position in the plane and pass the n small rings over this rod in such a way that the associated strings run parallel to the y-axis from the rings to the corresponding holes.

As in Sect. 7.2, the ith ring may be supposed to have moved a distance $|y_i - a - bx_i|$ from the ith hole at (x_i, y_i) to the point $(x_i, a + bx_i)$ on the rod, and thus the ith weight may be considered to have moved vertically through the same distance and thus gained potential energy proportional to $w_i|y_i - a - bx_i|$ so that the system as a whole has potential energy proportional to

$$\sum_{i=1}^{n} w_i|y_i - a - bx_i|.$$

When in equilibrium, the position of the rod will identify the values of a and b that define the L_1-norm fitted line.

Let $e_i = y_i - a - bx_i$ denote the ith residual, then the L_1-norm line fitting problem chooses a and b to minimise the sum of the weighted absolute residuals $\sum w_i|e_i|$. Denoting the scaled force in the ith string by s_i, we find that $s_i = -1$ if the ith residual e_i is negative, that $s_i = +1$ if e_i is positive and that s_i takes a value in the range $-1 \leq s_i \leq +1$ if e_i is zero.

Now, the force in the ith string is proportional to $w_i s_i$ and the rotational couple about the point $(0, a)$ on the rod is proportional to $w_i x_i s_i$. Further, the direct forces must sum to zero $\sum w_i s_i = 0$ as must the rotational couples $\sum w_i x_i s_i = 0$. These L_1-norm equilibrium conditions are to be compared with the more familiar least squares (or L_2-norm conditions $\sum w_i e_i = 0$ as $\sum w_i x_i e_i = 0$ outlined in Sect. 7.6 below.

It is to be noted that s_i does not represent the sign of e_i as s_i may take a nonzero value when e_i is zero. Readers who experience some difficulty in associating a horizontal force with a weight hanging vertically through the ith hole should consider the case of $n = 3$ observations with equal weights $w_1 = w_2 = w_3$. It may readily be established that the rod will usually pass through two of the points defining these holes. Further, since there must be a zero net force and a zero net couple acting on the rod when it is in equilibrium, we may deduce that, if the optimal position of the rod passes through two of the points defining these holes, then the weights hanging vertically through these holes must be associated with horizontal forces acting on the rod in such a way as to counterbalance the direct and rotational forces imposed on it by the third weight.

Newcomb (1873a, b) published two abstract models for the least squares (L_2-norm) variant of the unconstrained second and third problems in 1873, The first of these models is briefly described in Sect. 7.6 while the second (which is actually due to Donkin) serves as the basis of the L_1 -norm model described in Sect. 7.4. Readers are referred to Farebrother (1999) for a description of these L_2-norm models and for a complete transcription of Newcomb's first article. The explicit physical models for the constrained and unconstrained variants of the L_1-norm line fitting problem described in the present section are due to Farebrother (1987, 2002).

Making small changes to the weights in the model described above gives rise to the concept of L_1-norm influence. The corresponding least squares definition of this concept is to be found in Newcomb (1873a). And, more recently, the concept of L_1-norm influence was developed as the basis of a scheme of robust statistical analysis proposed by Hampel (1974) and Hampel et al. (1986).

By contrast with Newcomb (1873a) who varied a single weight and with Lamé and Clapeyron (1829) who were prepared to vary all n weights simultaneously, Koenker and Bassett (1978) were concerned with the effect of varying the weights associated with positive residuals differently from those associated with nonpositive residuals. Thus giving rise to the concept of regression quantiles whose use is described in several of the articles published in the volumes edited by Dodge (1987, 1992, 1997, 2002).

7.4 Fitting a Point to a Set of Lines in the Plane of Parameters: L_1-Norm Variants of Donkin's Problem

For our third problem, we again suppose that we have n pairs of observations on the variables X and Y which, as in Sect. 4.1, we choose to represent as n lines

$a = y_i - bx_i$ $(i = 1, 2, ..., n)$ in the ba-plane of parameters. Defining an additional point (b_0, a_0) in the same plane, we seek to determine the values of a_0 and b_0 which minimize the weighted sum of the absolute b-meridian distances

$$\sum_{i=1}^{n} w_i |a_0 - y_i + b_0 x_i|$$

between the arbitrary point and the n given lines where these b-meridian distances are measured parallel to the a-axis.

Once again, we may define a constrained variant of this third problem by noting that the elements a_0 and b_0 satisfy the condition $a_0 = y_0 - b_0 x_0$ if the arbitrary point (b_0, a_0) is to lie on the straight line $a = y_0 - bx_0$.

These two variants of our third problem represent the projective geometry duals of the constrained and unconstrained variants of the problem of Sect. 7.3, so that each of the n lines in the ba-plane corresponds to a single point in the xy-plane and *vice versa*.

In principle, it is also possible to develop the corresponding dual variants of the problem of Sect. 7.2, but, as noted in Sect. 7.1, it does not seem natural to fit a single line to a set of n lines in the ba-plane.

We may readily obtain a simple mechanical model for the line fitting problem in the space of parameters by passing a weighted string over each of the rigid rods representing the lines and attaching them to a ring at an arbitrary point in the ba-plane. If all of these strings are constrained to lie parallel to the a-axis, then, the ith weight clearly has potential energy proportional to $w_i |a_0 - y_i + b_0 x_i|$ so that the system as a whole has potential energy proportional to

$$\sum_{i=1}^{n} w_i |a_0 - y_i + b_0 x_i|.$$

As in Sect. 7.3, when the system is in equilibrium, the position of the ring will identify the values of a_0 and b_0 that define the arbitrary point (b_0, a_0) and hence the L_1-norm fitted line.

Now, the requirement that the strings in this model should lie parallel to the a-axis is inconvenient. To obtain a more natural alternative, we attach a weight of $v_i = w_i \sqrt{(1 + x_i^2)}$ at the lower end of the ith string which is unconstrained and thus permitted to take up a position at right angles to the ith rod. In this context, this revised third problem clearly minimises the v-weighted perpendicular distances from the arbitrary point to the n given lines

$$\sum_{i=1}^{n} v_i \left|(a_0 - y_i + b_0 x_i)/\sqrt{(1 + x_i^2)}\right|.$$

Thus, by the simple expedient of increasing the value of the weight attached to the lower end of the length of string passing over the rigid rod representing the ith line by a factor of $\sqrt{(1+x_i{}^2)}$ and by removing the constraint on the final position adopted by the strings of the model, we obtain an alternative, more natural, representation of the L_1-norm line fitting problem which compares favourably with that described above. When in equilibrium, the strings will lie at right angles to the associated rods and the position of the ring will identify the values of a_0 and b_0 that define the position of the L_1-norm fitted line.

Finally, we note that we have named this section for William Fishburn Donkin as he published an abstract least squares (L_2-norm) variant of the unconstrained third problem in 1844. Once again, the L_1-norm variant of this problem and the associated physical models are due to Farebrother (1987, 2002). However, it should also be noted that this model is essentially a generalisation of Varignon's model in which the pulleys are formally replaced by small rollers or frictionless rods.

7.5 Fitting a Point to a Set of Lines in the Plane of Observations: Oja's Bivariate Median

An interesting variant of the model of Sect. 7.4 occurs when we suppose that we are given m points $(x_i, \ y_i)$ $i = 1, 2, ..., m$ with distinct values of x_i, and that we are interested in fitting an arbitrary point $x_0, \ y_0)$ to these m observations in the following indirect manner. Taking these m points in pairs, we obtain a total of $n = m(m-1)/2$ lines with equations of the form

$$y = a_{ij} + xb_{ij}$$

where

$$a_{ij} = (x_j y_i - x_i y_j)/(x_j - x_i)$$

and

$$b_{ij} = (y_j - y_i)/(x_j - x_i).$$

Defining an arbitrary point $(x_0, \ y_0)$ in the xy- plane, we find that it is at a distance

$$e_{ij} = y_0 - a_{ij} - x_0 b_{ij}$$

from the ijth line when all distances are measured parallel to the y-axis, and at a distance

$$h_{ij} = e_{ij}/\sqrt{(1 + b_{ij}^2)}$$

when the ijth distance is measured perpendicular to the ijth line.

Further, on multiplying this ijth perpendicular distance by c_{ij}, the length of the line segment between (x_i, y_i) and (x_j, y_j), we have

$$d_{ij} = c_{ij}h_{ij} = w_{ij}e_{ij}$$

where $w_{ij} = |x_j - x_i|$ and

$$c_{ij} = \sqrt{[(x_i - x_j)^2 + (y_i - y_j)^2]}.$$

The absolute value of d_{ij} corresponds to twice the area of the triangle with vertices (x_i, y_i), (x_j, y_j) and (x_0, y_0). Whence we may deduce that Oja's (1983) bivariate median is formed by choosing values for x_0 and y_0 to minimise the sum of the areas of all n such triangles

$$\sum_{i<j} |d_{ij}| = \sum_{i<j} w_{ij}|e_{ij}|.$$

In passing, we note that this expression may also be written as:

$$\sum_{i<j} |d_{ij}| = \sum_{i<j} |(x_j - x_i)y_0 - (x_j y_i - x_i y_j) - (y_j - y_i)x_0|$$

where there is now no need to insist on distinct x-values.

To obtain a mechanical model for this procedure, we have to replace the m line segments connecting the m pairs of points in the xy-plane by rigid rods of arbitrary length, pass a length of string over the ijth rigid rod and attach a weight proportional to the length of the ijth line segment c_{ij} to the lower end of the ijth string and connect the upper ends to a ring in the xy-plane. When in equilibrium, the strings will lie at right angles to the associated rods and the position of the ring will identify the values of x_0 and y_0 that define Oja's bivariate median.

Now, by the rules of projective geometry duality outlined in Sect. 4.1, the m points and $n = m(m - 1)/2$ lines in the primal xy-plane correspond to a set of m lines with equations

$$a = y_i - bx_i$$

in the dual ba-plane. And the points of intersection of these m lines define the $n = m(m - 1)/2$ points (b_{ij}, a_{ij}) $i < j = 1, 2, ..., m$.

In this context, this problem takes the form of the weighted L_1-norm fitting of a straight line

$$a = y_0 - bx_0$$

to the grid of m points (b_{ij}, a_{ij}) $i < j = 1, 2, ..., n$ in the ba-plane using the objective function

$$\sum_{i<j} |d_{ij}| = \sum_{i<j} w_{ij}|e_{ij}|.$$

As in Sect. 7.3, we associate a real horizontal plane with the abstract ba-plane, then at each point (b_{ij}, a_{ij}) in this plane we drill a hole. Through this hole we pass a length of string with a weight w_{ij} at the lower end and a small ring at the upper end. We now place a rigid rod at an arbitrary position in the ba-plane and pass the m small rings over this rod in such a way that the associated strings run parallel to the a-axis from the rings to the holes. When this rod is in equilibrium, it again identifies the optimal values of x_0 and y_0 defining Oja's bivariate median.

For further details of the strings and pulleys models described in Sects. 7.2–7.4, see Farebrother (2002), and for further details of the models described in Sect. 7.5, see Farebrother (2006).

7.6 L_2-Norm Mechanical Models

In this Section and in the next we shall generalise the L_1-norm fitting problems outlined in Sects. 7.2 and 7.3 to the corresponding L_2-norm and L_∞-norm problems respectively. [It is also possible to generalise the problems of Sects. 7.4 and 7.5 in a similar way but these are left as exercises for the reader.]

In the first case, instead of choosing values for x_0 and y_0 to minimise the weighted sum of the absolute Euclidean distances from the n points to the arbitrary point

$$\sum_{i=1}^{n} w_i \sqrt{[(x_i - x_0)^2 + (y_i - y_0)^2]}$$

or values for the parameters a and b to minimise the weighted sum of the absolute x-meridian distances from the n points to the arbitrary line

$$\sum_{i=1}^{n} w_i |y_i - a - bx_i|$$

we choose values for x_0 and y_0 to minimise the weighted sum of the squared Euclidean distances from the n points to the arbitrary point

$$\sum_{i=1}^{n} w_i [(x_i - x_0)^2 + (y_i - y_0)^2]$$

or values for the parameters a and b to minimise the weighted sum of the squared x-meridian distances from the n points to the arbitrary line

$$\sum_{i=1}^{n} w_i [y_i - a - bx_i]^2.$$

In order to develop mechanical models for these weighted least squared deviation fitting problems, we install a second horizontal plane at unit distance below the first. A hole is drilled through the upper horizontal plane at the point indicated by the ith observations on the variables X and Y. A spring of unit natural length and modulus w_i (or, equivalently, a set of w_i springs of unit natural length and unit modulus) is passed through the ith hole and its lower end attached to the corresponding point on the lower horizontal plane. Then, the upper end of this ith spring is either tied to a ring lying at an arbitrary point in the upper horizontal plane or it is constrained to lie parallel to the y-axis before being tied to a ring which is passed over a rigid rod lying in an arbitrary position in the upper horizontal plane.

Now, the potential energy in a stretched spring of unit modulus is proportional to the square of its extension, so that the potential energy of the system as a whole is proportional to the weighted sum of the squared Euclidean distances from the n points to the arbitrary point

$$\sum_{i=1}^{n} w_i [(x_i - x_0)^2 + (y_i - y_0)^2]$$

or proportional to the weighted sum of the squared x-meridian distances from the n points to the rod

$$\sum_{i=1}^{n} w_i [y_i - a - bx_i]^2.$$

Further, the potential energy contributed by the ith spring

$$w_i [(x_i - x_0)^2 + (y_i - y_0)^2]$$

corresponds to a force in the ith spring proportional to

$$w_i \sqrt{[(x_i - x_0)^2 + (y_i - y_0)^2]}.$$

And, resolving this ith force parallel to the x-axis, we have $w_i(x_i - x_0)$; and, resolving it parallel to the y-axis, we have $w_i(y_i - y_0)$. Now, when the system is in equilibrium, these resolved forces must sum to zero, so that $\sum w_i(x_i - x_0) = 0$ and $\sum w_i(y_i - y_0) = 0$. Thus, our simple mechanical model identifies the weighted arithmetic means $x_0 = \sum w_i x_i / \sum w_i$ and $y_0 = \sum w_i y_i / \sum w_i$ as the optimal values of x_0 and y_0 in this case.

We now turn our attention to the familiar L_2-norm line fitting problem. Let $e_i = y_i - a - bx_i$ denote the ith residual, then the L_2-norm line fitting problem chooses a and b to minimise the weighted sum of the squared residuals $\sum w_i e_i^2$. In this case, the force in the ith spring is proportional to $w_i e_i$ and the corresponding rotational couple about the point $(0, a)$ on the rod is proportional to $w_i x_i e_i$. Moreover, when the system is in equilibrium, these direct forces must sum to zero $\sum w_i e_i = 0$ as must the corresponding rotational couples $\sum w_i x_i e_i = 0$. Gilstein and Leamer (1983) have identified a simple technique for determining the set of values of a and b for which both of these conditions are satisfied with nonnegative values for the moduli w_1, w_2, ..., w_n. Indeed, in the special case when all springs have the same unit modulus $w_1 = w_2 = ... = w_n = 1$ then these conditions clearly yield the familiar optimality conditions for the simple least squares line fitting problem.

As mentioned in Sect. 7.3, the abstract models for the L_2-norm line fitting s problem are due to Newcomb, but the explicit physical models described here are due to Farebrother (1987, 1999, 2002). [Alternative analyses of the L_2-norm point and line fitting problems which make use of the differential calculus are described in Chap. 4 of Farebrother (2002).]

7.7 L_∞-Norm and LMS Mechanical Models

In a similar way, instead of developing mechanical models for weighted L_2-norm variants of the weighted L_1-norm problems of Sects. 7.2 and 7.3, we may consider the corresponding *unweighted* L_∞-norm procedures which choose values for x_0 and y_0 to minimise the largest absolute Euclidean distance from the n points to the arbitrary point

$$max_{i=1}^{n}\sqrt{[(x_i - x_0)^2 + (y_i - y_0)^2]}$$

or values for the parameters a and b to minimise the largest absolute x-meridian distance from the n points to the arbitrary line

$$max_{i=1}^{n}|y_i - a - bx_i|$$

where it is convenient to ignore the possibility of differently weighted observations in the present context.

For a mechanical model of these unweighted minimax absolute deviation fitting problems, we have to combine the strings under tension from the models of Sects. 7.2 –7.4 with the pair of horizontal planes from the models of Sect. 7.6. We therefore suppose that we are given a pair of horizontal planes, the upper being fixed in position whilst the lower is free to move in a vertical direction. Holes are drilled through the upper horizontal plane at the n points indicated by the observations on the variables X and Y. A piece of string of given length is passed through each of these holes and attached to the corresponding point on the lower horizontal plane. The upper ends of the ith strings is either tied to a ring lying at an arbitrary point in the upper horizontal plane or it is constrained to lie parallel to the y-axis before being tied to a ring which is passed over a rigid rod lying in an arbitrary position in the upper horizontal plane. As the lower (movable) plane is gradually lowered, the strings tighten until, in the limit, it is not possible to lower this plane any further. This situation clearly determines the position of the ring (or rod) which minimises the largest of the absolute deviations between the ring (or rod) and the n holes. Points associated with slack strings are clearly nearer to the ring (or rod) than those associated with taut strings, and all points associated with taut strings will be at the same distance from the ring (or rod). The maximum deviation is thus given by the common length of the taut strings lying in the upper horizontal plane.

Further, since the corresponding least median of squares (LMS) procedures may be regarded as variants of the L_∞-norm procedures applied to a set of m observations (where m is chosen close to $n/2$), these mechanical models may readily be generalised to the corresponding least median of squares problems by selecting the relevant set of m observations (although, of course, this model is not able to explain the particular choice of m observations).

The explicit physical models for the L_∞-norm point and line fitting problem described here are again due to Farebrother (1987, 2002). Geometrical alternatives to these mechanical models are to be found in Chaps. 6 and 7 of Farebrother (2002). In particular, we note that the L_∞-norm problems are associated with three familiar geometrical instruments: a pair of callipers (or pincers) for determining the line segment of minimal length containing all n observations on a one-dimensional straight line, a pair of compasses for determining the circle of minimum radius containing all n observations on a two-dimensional plane, and a pair of parallel rules for determining the pair of parallel lines with minimal distance between them (measured parallel to the y-axis) containing all n observations on a two-dimensional plane.

References

Dodge, Y. (Ed.). (1987). *Statistical Data Analysis Based on the L_1-Norm and Related Methods*. Amsterdam: North-Holland Publishing Company.

Dodge, Y. (Ed.). (1992). *L_1-Statistical Analysis and Related Methods*. Amsterdam: North-Holland Publishing Company.

Dodge, Y. (Ed.). (1997). *L_1-Statistical Procedures and Related Topics*. Hayward: Institute of Mathematical Statistics.

Dodge, Y. (Ed.). (2002). *Statistical Data Analysis based on the L_1-Norm and Related Methods*. Basel: Birkhäuser Publishing.

Eisenhart, C. (1961). Boscovich and the combination of observations. In L. L. Whyte (Ed.), Roger Joseph Boscovich S.J. F.R.S., George Allen and Unwin (pp. 200–212). New York: London and Fordham University Press.

Farebrother, R. W. (1985). Linear unbiased approximators of the disturbances in the standard linear model. *Linear Algebra and Its Applications*, *67*, 259–274.

Farebrother, R. W. (1987). Mechanical representations of the L_1 and L_2 estimation problems, in Dodge (1987, pp. 455–464).

Farebrother, R. W. (1988a). *Linear Least Squares Computations*. New York: Marcel Dekker.

Farebrother, R. W. (1990). Further details of contacts between Boscovich and Simpson in June 1760. *Biometrika*, *77*, 397–400.

Farebrother, R. W. (1993). Boscovich's method for correcting discordant observations. In P. Bursill-Hall (Ed.), R. J. Boscovich: *His Life and Scientific Work* (pp. 255–261). Rome: Istituto della Enciclopedia Italiana.

Farebrother, R. W. (1999). *Fitting Linear Relationships: A History of the Calculus of Observations 1750–1900*. New York: Springer-Verlag.

Farebrother, R. W. (2002). *Visualizing Statistical Models and Concepts*. New York: Marcel Dekker. Available online from TaylorandFrancis.com.

Farebrother, R. W. (2006). Mechanical representations of Oja's spatial median. *Student*, *5*, 299–301.

Franksen, O. I., & Grattan-Guinness, I. (1989). The earliest contribution to location theory? Spatio-economic equilibrium with Lamé and Clapeyron 1829. *Mathematics and Computers in Simulation*, *31*, 195–220.

Gilstein, C. Z., & Leamer, E. E. (1983). The set of weighted regression estimates. *Journal of the American Statistical Association*, *78*, 942–948.

Hampel, F. R. (1974). The influence curve and its role in robust estimation. *Journal of the American Statistical Association*, *69*, 383–393.

Hampel, F. R., Ronchetti, E. M., Rousseeuw, P. J., & Stahel, W. A. (1986). *Robust Statistics : The Approach Based on Influence Functions*. New York: Wiley.

Koenker, R., & Bassett, G. W. (1978). Regression quantiles. *Econometrica*, *46*, 33–50.

Kuhn, H. W. (1967). On a pair of dual nonlinear programs. In J. Abadie (Ed.), *Methods of Nonlinear Programming* (pp. 29–54). Amsterdam: North-Holland.

Kuhn, H. W. (1973). A note on Fermat's problem. *Mathematical Programming*, *4*, 98–107.

Laplace, P. S. (1793). Sur quelques points du système du monde, Mémoires de l'Académie Royale des Sciences de Paris [pour 1789], pp. 1–87. (Reprinted in his Oeuvres Complètes, Vol. 11, Gauthier-Villars, Paris, 1895, pp. 447–558).

Laplace, P. S. (1799). Traité de Mécanique Céleste, Tome II. Paris: J. B. M. Duprat. (Reprinted in his Oeuvres Complètes, Vol. 2, Imprimerie Royale, Paris, 1843 and Gauthier-Villars, Paris, 1878. English translation by N. Bowditch, Hillard, Gray, Little and Wilkins, Boston, 1832. Reprinted by Chelsea Publishing Company, New York, 1966).

Laplace, P. S. (1812). Théorie Analytique des Probabilités. Paris: Courcier. 1812. Third edition with an introduction and three supplements, Courcier, Paris, 1820. (Reprinted in his Oeuvres Complètes, Vol. 7, Imprimerie Royale, Paris, 1847 and Gauthier-Villars, Paris, 1886).

Laplace, P. S. (1818). Deuxième Supplément to Laplace (1812).

Lamé, G., & Clapeyron, B. P. E. (1829) Mémoire sur l'application de la statique à la solution des problèmes relatifs à la théorie des moindres distances. Journal des Voies et Communications, 10, 26–49. (English translation by Franksen and Grattan-Guinness (1989, pp. 211–218)).

Newcomb, S. (1873a). A mechanical representation of a familiar problem. Monthly Notices of the Royal Astronomical Society, 33, 573. (Reprinted in Farebrother (1999, pp. 168–169)).

Newcomb, S. (1873b). Note on a mechanical representation of some cases in the method of least squares. *Monthly Notices of the Royal Astronomical Society*, *33*, 574.

Oja, H. (1983). Descriptive statistics for multivariate distributions. *Statistics and Probability Letters*, *1*, 327–332.

Stigler, S. M. (1984). Boscovich, Simpson and a 1760 manuscript note on fitting a linear relation. *Biometrika*, *71*, 615–620.

Stigler, S. M. (1986). *The History of Statistics: The Measurement of Uncertainty before 1900*. Cambridge: Harvard University Press.

Author Index

R. W. Farebrother, *L₁-Norm and L∞-Norm Estimation*,
SpringerBriefs in Statistics, DOI: 10.1007/978-3-642-36300-9,
© The Author(s) 2013